I0055800

A Course in Homogenization-Based Techniques

Multiscale Modeling and Asymptotic Analysis

Advanced Textbooks in Mathematics

Print ISSN: 2059-769X
Online ISSN: 2059-7703

The *Advanced Textbooks in Mathematics* explores important topics for post-graduate students in pure and applied mathematics. Subjects covered within this textbook series cover key fields which appear on MSc, MRes, PhD and other multidisciplinary postgraduate courses which involve mathematics.

Written by senior academics and lecturers recognised for their teaching skills, these textbooks offer a precise, introductory approach to advanced mathematical theories and concepts, including probability theory, statistics and computational methods.

Published

A Course in Homogenization-Based Techniques: Multiscale Modeling and Asymptotic Analysis
by Adrian Muntean

Geometry of Mechanics
by Miguel C Muñoz-Lecanda and Narciso Román-Roy

Classical and Modern Optimization
by Guillaume Carlier

An Introduction to Machine Learning in Quantitative Finance
by Hao Ni, Xin Dong, Jinsong Zheng and Guangxi Yu

Conformal Maps and Geometry
by Dmitry Beliaev

Crowds in Equations: An Introduction to the Microscopic Modeling of Crowds
by Bertrand Maury and Sylvain Faure

Mathematics of Planet Earth: A Primer
by Jochen Bröcker, Ben Calderhead, Davoud Cheraghi, Colin Cotter, Darryl Holm, Tobias Kuna, Beatrice Pelloni, Ted Shepherd and Hilary Weller
edited by Dan Crisan

More information on this series can also be found at
https://www.worldscientific.com/series/atim

Advanced Textbooks in Mathematics

A Course in Homogenization-Based Techniques

Multiscale Modeling and Asymptotic Analysis

Adrian Muntean

Karlstad University, Sweden

World Scientific

NEW JERSEY • LONDON • SINGAPORE • BEIJING • SHANGHAI • HONG KONG • TAIPEI • CHENNAI • TOKYO

Published by

World Scientific Publishing Europe Ltd.
57 Shelton Street, Covent Garden, London WC2H 9HE
Head office: 5 Toh Tuck Link, Singapore 596224
USA office: 27 Warren Street, Suite 401-402, Hackensack, NJ 07601

Library of Congress Control Number: 2025033015

British Library Cataloguing-in-Publication Data
A catalogue record for this book is available from the British Library.

Advanced Textbooks in Mathematics
A COURSE IN HOMOGENIZATION-BASED TECHNIQUES
Multiscale Modeling and Asymptotic Analysis

Copyright © 2026 by Adrian Muntean

All rights reserved.

ISBN 978-1-80061-829-9 (hardcover)
ISBN 978-1-80061-830-5 (ebook for institutions)
ISBN 978-1-80061-831-2 (ebook for individuals)

For any available supplementary material, please visit
https://www.worldscientific.com/worldscibooks/10.1142/Q0538#t=suppl

Desk Editors: Eshak Nabi Akbar Ali/Cian Sacker Ooi

Typeset by Stallion Press
Email: enquiries@stallionpress.com

To Andrea, Clara, Felix, and Victor

Preface

The world around us is often described by stationary and non-stationary mathematical models, which, though abstract, can offer surprising insights into complex phenomena. In many cases, these models can be computed numerically to make quantitative predictions, as long as certain assumptions hold true. A key example is the transport of matter through reactive porous media. With a bit of creativity, this example can be extended to composite materials with active microstructures, which are also highly relevant to the preservation of cultural heritage.

The primary message of this textbook is that asymptotic analysis, specifically homogenization-based techniques exploiting relevant examples, can offer powerful approximate models and concrete expressions for constitutive laws that bridge the gap between microscopic and macroscopic scales for a large variety of real-world scenarios, whose descriptions in terms of balance laws of extensive physical quantities are understood. On the practical side, understanding how to correctly couple these scales can unlock new interpretations of otherwise inaccessible microscopic data by linking them to macroscopic experimental observations. This work does not aim to cover the full range of steps typically involved in such analysis – steps that might include: (1) modeling the microscopic situation with balance laws, (2) scaling them properly, (3) developing a theory of well-posedness for the microscopic problem, (4) performing asymptotic analysis or upscaling, (5) establishing a well-posed theory for the upscaled problem, (6) justifying the asymptotics through corrector analysis, (7) designing convergent approximation schemes, (8) comparing upscaled models to experimental data, and (9) solving multiscale inverse problems. Instead, the textbook focuses on providing practical insights into selected mathematical techniques for multiscale problems, with a close eye on real-world applications.

The examples and exercises presented here should be seen as tools to illustrate the underlying theory – much like appetizers meant to spark further curiosity. These examples draw from classical mathematical questions as well as a few of my own research endeavors. While some are fully solved, others remain open, and many are part of established "homogenization folklore". Though I have made every effort to credit the original sources, I acknowledge that I may not always have succeeded in doing so completely.

The textbook is primarily aimed at graduate students in applied mathematics. Some chapters can serve as course material for first-year master's students, while others are suited for PhD students in applied analysis, computational science, and engineering. The goal is to equip readers with the tools of modern asymptotic analysis, empowering them to frame and address their own research questions.

As I continue to refine this work, I welcome feedback on any mathematical errors, omissions, suggestions for improvement, or pedagogical insights on how to effectively teach this content. Please feel free to reach out to me at adrian.muntean@kau.se if you have any comments or insights to share.

Adrian Muntean
Karlstad
June 17, 2025

About the Author

Adrian Muntean is a professor of mathematics at Karlstad University, Sweden. He teaches, supervises, and conducts research at the intersection of pure and applied mathematics. His research centers on mathematical modeling using differential equations and interacting particle systems, emphasizing their applications to real-world problems. Adrian's work spans diverse fields – including civil engineering, logistics, combustion engineering, soft matter, and complex systems – where he has advanced both understanding and practice. He is particularly recognized for his contributions to asymptotic methods for differential equations, with a specialized focus on the theory of homogenization.

Acknowledgments

I am deeply grateful to many individuals who have played a role in the creation of this textbook. I have learned the periodic homogenization technique through the invaluable guidance of colleagues and friends who generously shared their expertise. My training in this field began in 2005 at the Center for Industrial Mathematics in Bremen, Germany, during my collaborations with Sebastian Meier and Malte Peter, as well as Michael Böhm, who was our group leader at the time. This journey continued through numerous discussions with Iuliu Sorin Pop, Maria Neuss-Radu, and, last but certainly not least, Willi Jäger, whose influence as a scientist had a profound impact on me.

This textbook evolved from a set of lecture notes written with Vladimir Chalupecky several years ago, at the suggestion of Masato Kimura, for an intensive graduate course at the Institute of Mathematics for Industry at Kyushu University, Japan. It also draws from my regular homogenization course at Eindhoven University of Technology (2007–2015, jointly with Iuliu Sorin Pop) and at Karlstad University, Sweden (during the last 10 years, MAAD28). Special thanks go to my former students for their insightful comments and constructive criticism, which have been invaluable to this work.

I would like to extend special thanks to my collaborators who gave me feedback during the past few years. In particular, I am grateful to Kazunori Matsui and Erik Lieback for their thorough solutions to selected homework exercises and to Michael Eden, Rainey Lyons, Grigor Nika, Surendra Nepal, Vishnu Raveendran, and Arthur Vromans for their invaluable comments on the final version of this text. The chapter on finite element methods (Chapter 6) would not have been possible without Nicklas Jävergård, who patiently explained the structure of his FEniCS code. I am also deeply grateful to Kharisma Surya Putri for sharing her FreeFEM++ code, which provided alternative solutions to problems.

Rosie Williamson and Cian Sacker Ooi were the production editors from World Scientific Publishing responsible for this project. I thank you both for the professional assistance. Rosie deserves the credit for giving me the right push at the right moment to bring this textbook to a close.

Particularly, I am deeply grateful to my family – Andrea, Clara, Felix, and Victor, and it is truly hard to find the right words to express this feeling clearly. Your unconditional support and understanding made the completion of this manuscript possible. I apologize for my crazily frequent absences from our daily family life due to my many mathematical commitments. Your encouragement kept me going through every stage of this journey. Additionally, Clara, Felix, and Andrea have generously provided me with numerous illustrations for my text. Thank you for this!

Contents

List of Figures

Chapter 1

Introduction

In this chapter, we briefly present the main ideas behind the homogenization technique and anticipate potential applications to problem settings arising from materials science and life science, but not exclusively. Concepts such as multiple space scales, upscaling procedure, and simplified representations of heterogeneous media are introduced progressively.

1.1 Background

Multiscale modeling is today a modern topic in applied mathematics. It has direct applications in understanding complex processes of interest for materials and life science. For instance, in the study of composite materials (such as concrete members and paperboard) or biological membranes (soft matter in general), essential properties concerning transport and storage can rarely be understood by examining the associated physics at one length scale and one time scale alone. Furthermore, many potentially relevant length and time scales are prohibitively difficult to observe via experiments. So, what can be done to access the desired but unobservable information? Mathematical modeling offers a number of efficient techniques that can handle relevant multiscale problems, a joint effort in natural and computational sciences. In order to be useful, mathematical technology needs to be accessible to those involved (engineers, physicists, chemists, applied mathematicians, and so on).

My aim with this textbook is twofold: to train master's students in applied mathematics with techniques that are relevant for their education and could also be of practical importance in their careers (especially if they stay in academia or join R&D environments in industry), and to enhance the general mathematical literacy of scientists in the natural sciences and engineering.

1.2 Why is "Multiscale" Interesting for CSE?

Depending very much on the concrete application at hand, it is often hard to identify which parts of mathematics are involved in each context. However, concerning multiscale modeling in the playground of partial differential equations, one can expect that asymptotic methods for differential equations will combine with techniques specific to partial differential equations, applied functional analysis, numerical analysis, scientific computing, data assimilation, parameter estimation, and data visualization. If interacting particle systems intervene and play a part in the modeling, then statistical physics, stochastic processes, and statistical and machine learning become naturally involved. To deal with situations with increased complexity, a new science track is formed that bridges across multiple disciplines – the so-called *Computational Science and Engineering* (CSE). Homogenization-based techniques are mathematical methodologies for multiscale problems that belong to the CSE toolbox.

The core topic of this textbook – the periodic homogenization technique – approaches CSE from the perspective of asymptotic methods in partial differential equations. Essentially, one wants to understand complex multiscale problems (arising in composite materials or, more generally, in porous media). The starting point is normally a set of balance equations (partial differential equations) posed in a *heterogeneous medium*, say Ω, a non-empty set in $\mathbb{R}^d$ ($d \in \{1, 2, 3\}$). A typical example of such a set Ω is described in Figure 1.1 (right-hand side). Within this context, the word "heterogeneous" refers to a medium that has well-defined internal structures. It can be made of a single phase (for example, turbulence in air or wave structure of water surface) or of multiple phases (like a mix of solid, gas, and/or liquid phases present all together at the pore level in a porous medium). In Figure 1.1 (left), the set Y_x^s represents a solid grain, while Y_x^l indicates a liquid film attached to it. The set $Y \setminus (Y_x^s \cup Y_x^l)$ indicates the free space occupied by the gas phase. In porous media terminology, this is called "pore space". We refer to such an internal structure as *microstructure*. For instance, each droplet in Figure 1.2 or any black square on the chessboard shown in Figure 1.4 can be seen as a microstructure. In the same spirit, the chessboard (perceived as a whole) and the picture shown

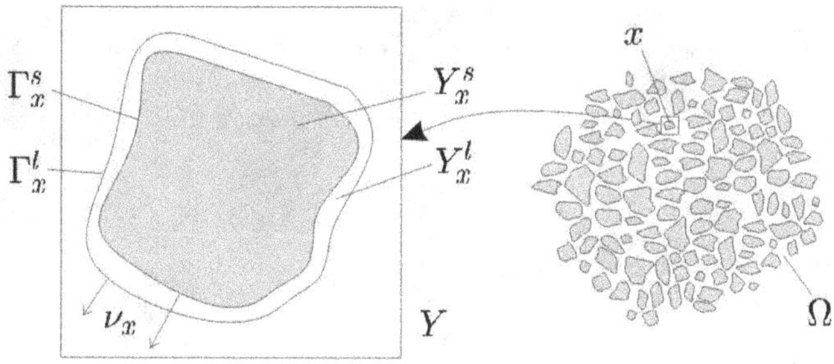

Figure 1.1. Example of an heterogeneous domain with distributed microstructures.

in Figure 1.1 (right) are typical representations of porous media. For further descriptions of porous media, we refer the reader to Bear (1988).

In Figures 1.1 (right-hand side) and 1.2, we see that microstructures are typically dependent on their spatial position in Ω, i.e. droplets are not identical. We label such microstructures as being x-dependent, with $x \in \Omega$. They can also depend on the time variable, say $t \in [0, T]$ for some $T > 0$. In this case, we label them additionally as being t-dependent. It can also happen that microstructures vary in both time and space variables. This is the case when the microscopic problem is a free- or moving-boundary problem. If the microstructures are identical, as with the chessboard, and do not change their shape as time elapses, then they build a periodic tiling of the domain. We speak in such a case about a "periodic microstructure", that is, the domain Ω is covered by an array of periodically repeated microstructures. The case of periodic microstructures is perhaps the most interesting one, as often the performed asymptotic analysis can be made mathematically rigorous, and hence, the result of the homogenization (averaging or upscaling) procedure can be fully trusted. Occasionally, we give hints on what can be technically done when the periodicity assumption is relaxed, as in the locally periodic case (see Figure 1.2) or the random case (see Figure 1.3).

As a research field, CSE is mostly focused around performing fast and accurate computations on advanced mathematical descriptions (evolution equations) of strongly coupled multi-physics scenarios formulated for heterogeneous materials. Interestingly, even if model equations were decoupled and linear, performing fast and accurate computations on complex geometrical structures is often out of reach. The reason is simple: Too many details are available. Hence, one encounters tremendous difficulties in

Figure 1.2. A nearly periodic arrangement of rain droplets on my car. If such an arrangement of droplets can be obtained via a small perturbation of a periodic configuration, then we will refer to it as a locally periodic arrangement.

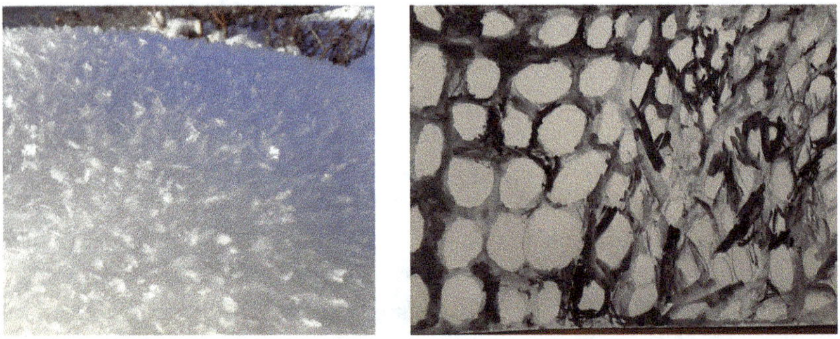

Figure 1.3. Left: Randomly oriented hoarfrost crystals. How far is this randomness from a periodic arrangement? Right: Snowing. Courtesy of Clara Muntean (when she was seven years old).

distinguishing computationally between what is relevant and what is not. In a somewhat similar way, disordered systems are impossible to handle to ensure that relevant information is captured. Mathematical homogenization tools can be of help if some good simplifying feature is present in the geometry of the domain. The periodicity of the microstructure is such a feature.

However, high oscillations are expected to occur when evolution equations are posed on arrays of periodically repeated microstructures. The good news is that, precisely in this case, one can upscale reasonably well and get a good grip on upscaled equations and effective coefficients so that the mean-field behavior is well described. Sometimes, a part of the information hidden at the level of fluctuations can also be traced in terms of controlled *a priori* bounds called "correctors" (or convergence rates) of the homogenization limiting process. Additionally, we note that in recent years, the universal approximation property of neuronal networks has been used as an alternative method to approximate numerically various types of mathematical models formulated in terms of partial differential equations. Consequently, conceptually new developments have arisen where periodic homogenization techniques are combined with machine learning algorithms to get a better grip on the role played by the microstructures in the averaging process; see e.g. Soyarslan and Pradas (2024). However, as the underlying theory is not yet ripe, we are not discussing in this textbook connections between neuronal network representations and averaging techniques. We anticipate though that, in the coming years, new scientific discoveries will happen in this direction.

1.2.1 *Toward upscaling information on periodic structures*

Depending on the level of description, one or more upscaling methods may be available. Most of the approaches rely on formal asymptotic arguments. Some of them can be justified rigorously using functional analytic techniques. This is really great news, as mathematics guarantees the trust in the obtained result. We enumerate here a few possibilities and also indicate a relevant bibliographic reference where more information can be found on the particular subject. Our list is not exhaustive.

For instance, if one has in mind capturing transitions between phenomena occurring at continuum microscopic scales and others at continuum macroscopic scales, then upscaling techniques include mixture theory (Bowen, 1980), volume averaging (Davit *et al.*, 2013), renormalization group methods (Braga *et al.*, 2003), matched asymptotics (Chapman *et al.*, 2015) (cf. also Sections 11 and 12 in Chapter 2 in Chechkin *et al.* (2007)), or asymptotic homogenization for periodic, locally periodic, and weakly random structures (Le Bris, 2015). On the other hand, one is often interested in understanding the discrete and/or the stochastic origins of phenomena observed macroscopically at a continuum scale. In such cases, upscaling techniques include statistical mechanics (mostly for equilibrium scenarios) with mean-field and hydrodynamic limits (Presutti, 2009), discrete-to-continuum limits (van Meurs *et al.*, 2014), and coarse graining (E., 2011).

These techniques can be framed into a larger class, which we refer to as averaging techniques.[1] Depending on the problem at hand, some of the averaging methods can outperform others. In general, a fair comparison between any two averaging methods is quite difficult to make; for some attempts in this direction, we refer the reader to Leguillon (1997) and Davit *et al.* (2013). If, for instance, microstructures enjoy periodicity[2] properties, then a few of the upscaling techniques have a good chance of being justified mathematically, and hence they can be trusted in applications. This is the case with the two-scale asymptotic homogenization method, which is the focus of this text.

1.2.2 *Heterogeneous media and their homogeneous representation*

Let us begin with a *Gedankenexperiment.* Imagine for a moment a material resembling a chessboard. Assume that the white and black squares are made of very different materials, each of them being homogeneous. In particular, say that these two materials have a uniform but different response to heat conduction, i.e. they have two distinct heat conductivities. Assume that the white square is made of Cheddar cheese, while the black one is made of steel. We call such an imaginary composite material *cheeso-steel.* Note that the heat conductivity for steel is around 50.24 W/mK, while for Cheddar cheese it is around 0.34 W/mK, cf. Marschoun *et al.* (2001). A more "orthodox" composite would be, for instance, the one obtained by combining bismuth (heat conductivity of 9.7 W/mK) and copper (heat conductivity of 413 W/mK). Observe in both cases the rather high difference in the conductivity values. In mathematical terms, it means that jumps will appear in the gradients of the temperature each time a black–white interface is crossed.

---

[1] This is a personal viewpoint. Other researchers have a slightly different perspective on the used terminology, e.g. Pavliotis and Stuart (2008) make a fine distinction between concepts such as averaging and homogenization.

[2] As an alternative to the desired periodicity, some other type of geometric information could be equally useful, as in the case of fractal sets or when the microstructure matches the ergodicity and stationarity properties of the underlying measure.

Solving numerically a heat conduction equation for a white (or black) square is elementary.[3] Also, solving the same equation for the cheeso-steel board (8×8 squares) is elementary when perfect transmission conditions are prescribed at all black–white interfaces. Repeating the experiment for an increasing number of squares, one observes immediately that oscillations start to occur in the numerical approximation of the temperature field. These oscillations become even stronger when approximating the temperature gradient. Consequently, naive numerical discretization methods supporting a direct computational approach are prone to fail. This observation makes us wonder whether it would be possible to solve a somewhat modified heat conduction equation posed on a homogeneous (gray) board instead (see Figure 1.4 (right)) and obtain for that case easily numerical results that are close, in suitable norms, to the solution of the original oscillating problem.[4] This would completely avoid the occurrence of oscillations, introducing instead a (hopefully small) modeling error.

We do not know yet what this modified heat conduction equation should look like, but if we did know it, then approximating numerically in an efficient way heat conduction problems formulated for composites with high

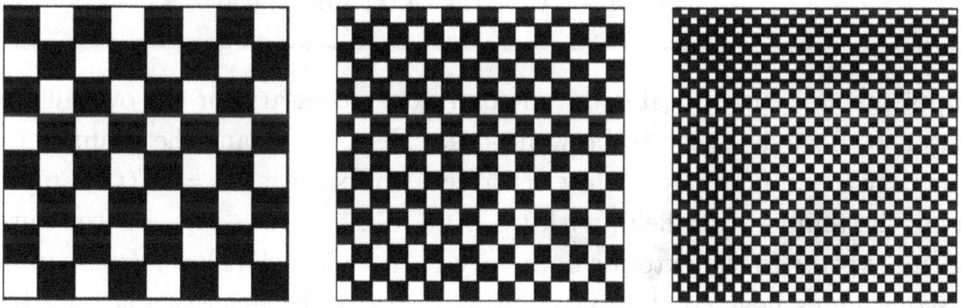

Figure 1.4. Chessboard-like geometries.

[3]Play the game by trying to solve numerically the imposed scenario, for instance, in FEniCS (Alnaes *et al.*, 2015) (or in any other open-source computational platform you are familiar with). Repeat your numerical experiment when white and black squares are linked in a chessboard. Increase the number of squares from 64 to 256, while keeping the length of the chessboard side unchanged. Now, repeat the computations.

[4]As when translating texts from one language into another (like from Japanese into English): "Is it better to translate them literally so as not to 'betray' the text but at the risk of a lower-quality translation? Or, is it perhaps better to find the closest alternative that makes sense in the target language, even though the translated version may slightly modify the idea?"

contrasting inclusions[5] becomes possible. Together with Auriault (1991), we ask the following: "When is it possible to formulate an equivalent problem in a homogeneous domain?" To address the question, we make use of weak convergence techniques for partial differential equations (Evans, 2015), combined with formal asymptotic arguments.

Anticipating the main idea behind the homogenization method, imagine for a moment that the following boundary-value problem describes the transfer of heat through our cheeso-steel board, which we denote by Ω_ε. Here, $\varepsilon > 0$ is a small dimensionless parameter that compares two characteristic length scales as follows: $\varepsilon := \frac{\ell_{micro}}{\ell_{macro}}$. The microscopic length scale ℓ_{micro} pinpoints the periodicity length, and the macroscopic length scale ℓ_{macro} takes into account the desired observation scale.[6] Let S be a time interval during which we observe the overall scenario. The target problem is as follows:

Find u_ε such that

$$\partial_t u_\varepsilon + \mathrm{div}(-A_\varepsilon(x)\nabla u_\varepsilon) = f_\varepsilon \text{ in } S \times \Omega_\varepsilon$$
$$[u_\varepsilon] = 0, \; [n \cdot A_\varepsilon \nabla u_\varepsilon] = 0 \text{ at black–white inner boundaries,}$$
$$-n \cdot A_\varepsilon \nabla u_\varepsilon = 0 \text{ at the outer boundary,}$$

supplemented with suitable initial conditions necessary for the overall evolution to have a chance to be well-posed. Here, $[\zeta]$ means the jump in the quantity ζ across a black–white interface. To fix ideas, $u_\varepsilon = u_\varepsilon(t, x)$ represents the temperature field, with $t \in S := (0, T)$ and $x \in \Omega_\varepsilon$, n is the outer normal vector to the exterior boundary of Ω_ε, while $A_\varepsilon(x) = A(x, \frac{x}{\varepsilon})$ is the discontinuous heat conductivity. Moreover, x is the macroscopic variable, and $\frac{x}{\varepsilon} =: y$ is the microscopic variable supposed to capture the level of the oscillations produced when varying ε. Furthermore, f_ε and g_ε are given oscillating data.

A first natural idea is to use the *ansatz*

$$u_\varepsilon(t, x) = u_0(t, x) + \varepsilon u_1\left(t, x, \frac{x}{\varepsilon}\right) + \varepsilon^2 u_2\left(t, x, \frac{x}{\varepsilon}\right) + \mathcal{O}(\varepsilon^3)$$

[5]In the heat conduction context, "high contrasting inclusions" refer to composite materials whose phases have very different heat conductivities, the difference between their values being sometimes of many orders of magnitude.

[6]Typically, one wants well-separated scales, i.e. $\ell_{micro} \ll \ell_{macro}$. What do you think would be good candidates for ℓ_{micro} and ℓ_{macro} in the chessboard?

to derive partial differential equations for u_0. Here, $u_\ell(x, \cdot)$ ($\ell \in \mathbb{N}$) are taken to be periodic functions (with the same period!) in the second variable. As this *ansatz* might not always hold, one has to come up with a rigorous justification of the proposed asymptotic behavior, i.e. one should prove a *corrector estimate* such as

$$||u_\varepsilon - u_0||_{\mathcal{X}} \leq c(u_1, \dots)\varepsilon^\alpha,$$

with $\alpha > 0$ and $c > 0$, with both constants being independent of the choice of ε. In this context, $\mathcal{X}$ is a suitable space where the two functions u_ε and u_0 can be compared with each other. Some researchers refer to this type of estimate as the convergence rate for the homogenization-limiting process. We like to call it the *corrector estimate*, as it provides information on how to correct the limit function u_0 with a term of the order of $\mathcal{O}(\varepsilon^\alpha)$. In the same spirit, corrector estimates provide a way to express in a quantitative way the range of validity of the upscaled (homogenized, macroscopic, averaged, etc.) model equations. In other words, a rigorous proof of corrector estimates justifies the validity of the homogenization-limiting procedure, i.e. it shows that the *ansatz* suggested earlier was in fact correct. Note that a direct consequence of the corrector estimate is the strong convergence

$$u_\varepsilon \to u_0 \quad \text{in } \mathcal{X} \text{ as } \varepsilon \to 0.$$

We now wonder what modified heat conduction equation should the limit function u_0 satisfy. At this moment, we can only speculate that the *upscaled* equation describing the conduction of heat through our cheeso-steel reads as follows:

Find $u_0 \in \mathcal{X}$ such that it holds

$$\partial_t(\phi u_0(t,x)) + \operatorname{div}_x(-\phi D(x, w(x))\nabla_x u_0(t,x)) = \phi f_0(t,x) \text{ in } S \times \Omega$$
$$-n \cdot \phi D(x, w(x))\nabla_x u_0(t,x) = 0 \text{ at the outer boundary,}$$

supplemented with corresponding initial conditions, where:

- ϕ is the *porosity* of the chesso-steel board;
- $D(\cdot, w(\cdot))$ is the *effective* heat conductivity coefficient;
- w is a function depending implicitly on the chessboard's microstructure, which will be referred to within this framework as the *cell function*.

If ϕ is the porosity of the chessboard, then this is a constant. On the other hand, if ϕ is the porosity of a material similar to the sketch shown in Figure 1.1, then we would have $\phi = \phi(x)$. If the "pores" host evolving-in-time free interfaces, then we expect $\phi = \phi(t, x)$, or even $\phi = \phi(t, x, u_0(t, x))$.

As corrector estimates, we expect the following inequalities to hold (with strictly positive constants c_1, c_2, c_3, T uniform in ε and $S := (0, T)$ is some observation time slot):

(i) $||u_\epsilon - u_0||_{L^2(S,\Omega_\varepsilon)} \leq c_1\varepsilon$;
(ii) $||\partial_t u_\varepsilon - \partial_t u_0||_{L^2(S,\Omega_\varepsilon)} \leq c_2\varepsilon$;
(iii) $||\nabla u_\varepsilon - \nabla_x u_0 - \text{oscillations}||_{L^2(S,\Omega_\varepsilon)} \leq c_3\sqrt{\varepsilon}$.

It is worth pointing out at this stage itself that u_ε and u_0 reside normally in different function spaces. Hence, if one wants to compare them, one must choose between considering them either at the microscopic level, Ω_ε, or at the macroscopic level, Ω. In the first case, one has to find out a way to project u_0 back at the level of Ω_ε, while in the second case, one has to imagine u_ε extended to the whole of Ω. To facilitate a comparison between macroscopic and microscopic quantities, the functions u_0, $\partial_t u_0$, and $\nabla_x u_0$ arising in (i)–(iii) above are reconstructed at the level of the microscopic domain Ω_ε.

Interestingly, one can prove rigorously that the effective conductivity of the cheeso-steel board is approximately 4.13 W/mK, which is precisely the geometric average between the conductivities of the two components of the mixture. We refer the reader to Proposition 2.16 in Le Bris (2015) for a rigorous treatment of an elliptic case closely related to the case of heat propagation through the cheeso-steel board discussed here.

1.2.3 *Mathematical multiscale methods and applications as work in applied analysis*

It is possible to bring multiscale mathematical techniques closer to real-world applications. By doing so, one can improve existing models, refine mathematical results, and shed light on unresolved laboratory experiments. Achieving this typically requires broad scientific expertise, and it certainly helps to be part of a highly interdisciplinary, research-driven team. Such an endeavor is rarely the product of a single individual, especially when addressing various aspects of a multiscale problem. This includes not only mathematical modeling but also rigorous asymptotic analysis of evolution equations, numerical approximation, the implementation of multiscale discretizations and algorithms, data visualization, and parameter identification via deterministic approaches or via statistical inference and/or machine learning.

The community focused on these research areas is vast, and the range of applications for upscaling techniques is truly impressive. To highlight just a few potential applications, we briefly mention a selection of topics that we have explored in collaboration with our colleagues, namely: homogenization of locally periodic materials (Muntean and van Noorden, 2013; Fatima *et al.*, 2011); sulfate attack in sewer pipes (Fatima and Muntean, 2014); smoldering combustion in porous media (Fatima *et al.*, 2014); corrector estimates for homogenization of thermoelasticity equations and of selected coupled reaction-diffusion systems (Muntean and van Noorden, 2013; Eden and Muntean, 2017; Vo Anh and Muntean, 2019); discrete-to-continuum limits for dislocation dynamics (van Meurs *et al.*, 2014); Monte Carlo dynamics of active interacting particles in the dark (Cirillo and Muntean, 2013); finite element method (FEM) approximating evolution equations posed on multiple scales (Muntean and Neuss-Radu, 2010; Lind *et al.*, 2020); analysis and approximation of measure-valued equations as applied to opinions and crowd dynamics (Evers *et al.*, 2015); dynamics of human crowds in heterogeneous domains in the presence of fire (Richardson *et al.*, 2019); homogenization of a pseudo-parabolic system arising in the sulfatation of concrete (Vromans *et al.*, 2019); models for active–passive pedestrian flows (Cirillo *et al.*, 2020); deposition, coagulation, and fragmentation of colloids in porous media (Muntean and Nikolopoulos, 2020); homogenization of particle flows through heterogeneous thin membranes (Raveendran *et al.*, 2022), and nonlinear dispersion as a consequence of fast drifts occurring in homogenization settings (Raveendran *et al.*, 2025), where micro–macro computations need to be done efficiently, often relying on high-performance computing strategies such as that reported by Lakkis *et al.* (2024).

In most of these examples of direct applications of the homogenization methodology, the heterogeneity of the domain is described using periodically perforated geometries; see e.g. the domains depicted in Figure 1.5. Intuitively, keeping the macroscopic length fixed, the number of small-volume inclusions increases as ε goes to zero. When ε reaches zero, the inclusions vanish away, leading to a homogeneous domain as the end result (inclusions become zero-sized but have a controlled number, which leads to a fixed and finite porosity).

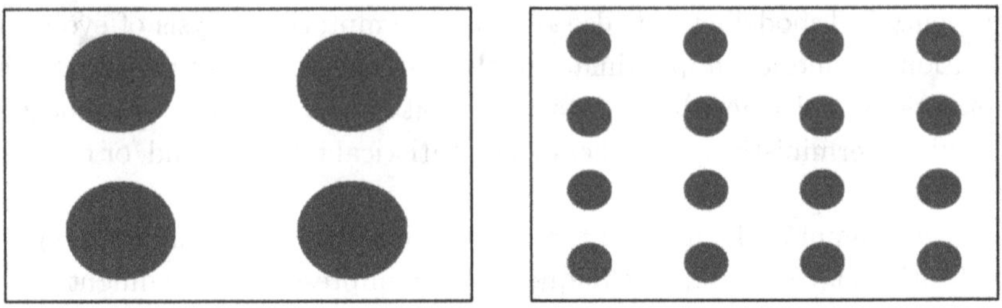

Figure 1.5. Sketch of two-dimensional perforated structures for two distinct choices of ε; The white and gray colors indicate two different materials.

1.3 A Note on Methodology

The structure of a one-semester course for master's students in applied mathematics could include the following sections:

(1) Introduction to Averaging Techniques (Sections 1.2.2, 2.1, and 7.4),
(2) Averaging Linear Elliptic Equations via Two-Scale Formal Homogenization Asymptotics (Section 2.3),
(3) Averaging Linear Elliptic Equations Using Two-Scale Convergence (Sections 7.3 and 7.6 and Chapter 3),
(4) Handling Deterministically Perforated Domains (Chapter 4),
(5) Further Exploration of Scaling and Correctors (Sections 3.2.2 and 4.6),
(6) Numerical Solutions for Microscopic and Macroscopic Models (Chapter 6),
(7) Homework Exercises, Additional Self-Study, Group Projects, and Examination.

Each chapter in this textbook includes a selection of homework exercises. Many of them are solved in detail, while others come with hints to guide the solution. The remaining exercises are intentionally left without any solution strategy.

1.4 Comments on the Literature

This textbook evolved from my previous lecture notes on homogenization (Muntean and Chalupecky, 2011) and Meier and Muntean (2010), as well as from the one-semester master's level courses on asymptotic analysis and periodic homogenization that I have taught over the years at Eindhoven University of Technology (the Netherlands), the Dutch Mastermath Program at Utrecht (the Netherlands), and Karlstad University (Sweden). Portions of this material have also been used multiple times for intensive training sessions with PhD students in applied mathematics.

The mini-courses (Allaire, 2002a; Alouges, 2016; Alexandrian, 2015) align closely with the spirit of this textbook. In the existing homogenization literature, the books by Le Bris (2015) and Berlyand and Rybalko (2018) were written with a similar goal in mind: to introduce the homogenization technique to a broad audience. For those interested in the practical aspects of the homogenization method and its way of thinking, we recommend further reading, particularly Cioránescu and Donato (1999), Pavliotis and Stuart (2008), Sanchez-Hubert and Sanchez-Palencia (1992, 1993), Mei and Vernescu (2010), and E. (2011). Advanced monographs on periodic homogenization include Bakhalov and Panasenko (1989), Bensoussan *et al.* (1978), Allaire (2002b), Marchenko and Khruslov (2006), and Chechkin *et al.* (2007).

The bibliography provided is by no means exhaustive, and many valuable resources may have been omitted. Additional references will be included in the upcoming chapters. Interested readers are encouraged to explore the literature further.

1.5 Exercises

Exercise 1.5.1. Draw two periodic pavements of $\mathbb{R}^2$ or $\mathbb{R}^3$ (also referred to as periodic tilings or periodic tessellations). In one drawing, the repeated pattern (we call it "microstructure") has a Lipschitz boundary, while in the other drawing the microstructure does not have a Lipschitz boundary. Estimate what boundary regularity you expect for your drawings.

Exercise 1.5.2. For $\varepsilon = \frac{1}{n}$ with $n \in \mathbb{N}$, consider the family of problems

$$\frac{d}{dx}\left(-a_\varepsilon(x)\frac{d}{dx}u_\varepsilon(x)\right) = 0 \quad \text{for } x \in (0,1),$$

$$u_\varepsilon(0) = 0, \quad u_\varepsilon(1) = 1,$$

where the coefficient $a_\varepsilon(x) := a(x/\varepsilon)$, with $a(\cdot)$ being a 1-periodic function, continuous, and bounded away from zero. Propose a constant $a^\star > 0$ such that the solution $u^\star$ of the problem

$$\frac{d}{dx}\left(-a^\star \frac{d}{dx}u^\star(x)\right) = 0 \quad \text{for } x \in (0,1),$$

$$u^\star(0) = 0, \quad u^\star(1) = 1$$

provides an $\mathcal{O}(\varepsilon)$ approximation (in the maximum norm) of u_ε.

Exercise 1.5.3. Take as $a(\cdot)$ your favorite periodic function scaled so that the period becomes 1 and make a choice also for the function $f : (0,1) \to (0,1)$. To keep things simple, take $f \in C([0,1]) \cap C^1((0,1))$, and set $\varepsilon > 0$ presumably small. Use `Python` to solve numerically the boundary-value problem discretized with finite differences of grid size $h > 0$: Find a function u_ε^h which, for small values of $h > 0$, is approximating sufficiently well the function u_ε solution to the boundary-value problem

$$\frac{d}{dx}\left(-a\left(\frac{x}{\varepsilon}\right)\frac{d}{dx}u_\varepsilon(x)\right) = f(x) \quad \text{for } x \in (0,1),$$

$$u_\varepsilon(0) = u_\varepsilon(1) = 0.$$

Consider the following cases:

(i) $h > \varepsilon$;
(ii) $0 < \varepsilon \ll h$.

Exercise 1.5.4. Take $b \in L^2(\mathbb{R})$ and the same hypothesis as in the previous exercise. Now, we are asking the same question as before, formulated for the following modified problem: Find a function u_ε^h which, for small values of $h > 0$, is approximating sufficiently well that function u_ε solution to the boundary-value problem

$$\frac{d}{dx}\left(-\left(a\left(\frac{x}{\varepsilon}\right) + b\left(\frac{x}{\varepsilon}\right)\right)\frac{d}{dx}u_\varepsilon(x)\right) = f(x) \quad \text{for } x \in (0,1)$$

$$u_\varepsilon(0) = u_\varepsilon(1) = 0$$

for the cases

(i) $h > \varepsilon$;
(ii) $0 < \varepsilon \ll h$.

1.6 Solutions

Solution 1.5.1. Before any drawing exercise, it makes sense to have at hand a standard book on function spaces to clarify the concept of domain and boundary regularity. The curious reader could have a look at the periodic tessellations of $\mathbb{R}^2$ by M. C. Escher; check either Wikipedia for "regular division of the plane drawings by Escher" or browse Escher's official collection https://mcescher.com/gallery/. We advise you to consider any of Escher's periodic tessellations and approximate locally the regularity of their boundary. Alternatively, pick an Indonesian batik motif from https://en.wikipedia.org/wiki/Batik. Such a motif typically combines periodic floral, animal, and geometrical designs. Check its boundary regularity. Justify your answer. Can you generate numerically your own batik-type pattern, eventually similar to the one shown in Figure 1.6?

As side information, we note that in plane geometry, the einstein problem asks about the existence of an aperiodic shape (prototile) that can tessellate space but only in a non-periodic way. This shape is called "einstein",

Figure 1.6. Example of a Python-generated batik-like motif.

i.e. “ein Stein”, meaning “one stone” in German. One can use such an einstein to construct a tessellation which has a Lipschitz boundary but fails to be periodic. Remotely connected to this discussion, the reader may note that they can pose the homogenization question for a large variety of tiles not necessarily building a periodic tessellation (such as Penrose tiles and Wang tiles). They are sometimes associated with “quasicrystals”.

Solution 1.5.2. This is one of the standard, exactly computable homogenization examples in one dimension. We refer the reader to Section 1.2.1 in Hornung (1997) for eventual hints. The conclusion will be that one can write down, for any $x \in (0,1)$, the following exact representation: $u_\varepsilon(x) = u^\star(x) + \epsilon u_1\left(x, \frac{x}{\varepsilon}\right)$, with computable functions $u^\star(\cdot)$ and $u_1\left(\cdot, \frac{\cdot}{\varepsilon}\right)$ for any $\varepsilon > 0$. Here, u_1 is a reminder function with an explicit construction.

Solution 1.5.3. We recommend the use of computational resources that are open source. For instance, `Python` is a very good option, while `Julia` is increasingly attractive, but other good alternatives exist as well. Feel free to choose whatever programming language or computational platform fits you best.

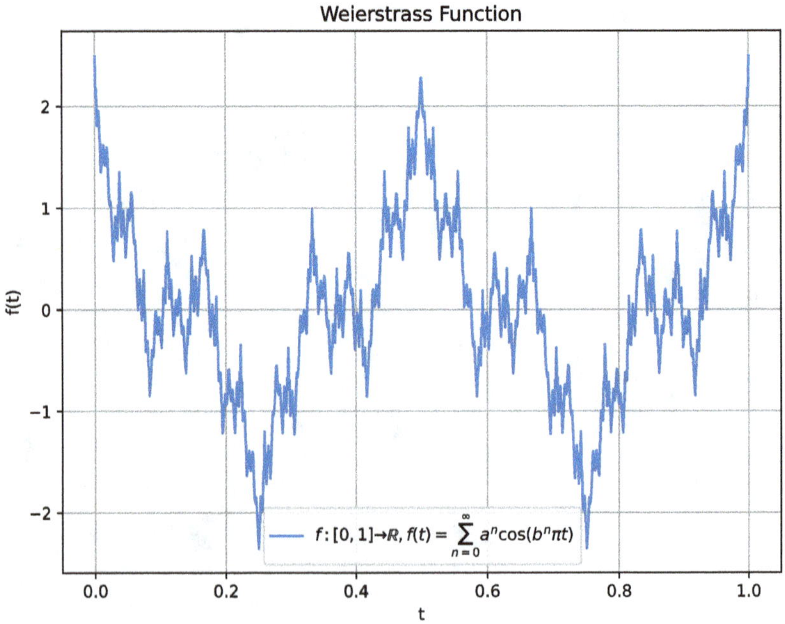

Figure 1.7. Plot of the Weierstrass function when the sum is truncated at $n = 200$, while the other parameters are $a = 0.6$ and $b = 3$.

To nicely illustrate numerically the captured solution behavior, one should be smart in picking the numerical values for h and ε. For this type of problem, it is always advisable to manufacture beforehand an exact solution and then try to approximate it numerically in an accurate fashion. As a side note, one may be tempted to require only $f \in C([0,1])$ and overlook additional smoothness. However, if one takes the Weierstrass function, i.e. $f : [0,1] \to \mathbb{R}$ defined by $f(t) := \sum_{n=0}^{\infty} a^n \cos(b^n \pi t)$ (with $0 < a < 1, b > 0$ an odd integer and $ab > 1 + \frac{3}{2}\pi$), then this exercise becomes very cumbersome, as the graph of this function is a fractal curve (see Figure 1.7), i.e. it is particularly hard to choose, for instance, the appropriate value for the grid size h.

By assuming the C^1-regularity for the function $f(\cdot)$, we discard such a pathological case.

Solution 1.5.4. Observe that $b(\cdot)$ is here some sort of defect from periodicity (especially if it is strongly localized). It is meaningful to compare the situation when $b(\cdot)$ is identically zero with the one when it is not vanishing everywhere. A quite inspiring reading in the context of periodic homogenization problems with deterministic defects is the work of Blanc *et al.* (2012).

Chapter 2

First Homogenization Attempts: Rigorous Convergence Arguments and Formal Two-Scale Asymptotics

In this chapter, we combine rigorous and formal arguments to perform a basic upscaling strategy that relies on the periodicity of the coefficients in model equations. The technique is usually called periodic asymptotic homogenization. We refer to it simply as "homogenization". In Section 2.1, we are concerned with the homogenization of a system of linearly coupled second-order (elliptic) equations posed in one dimension (1D). In Section 2.2, we perform the homogenization of a linear second-order elliptic equation for the simplest yet relevant two-dimensional (2D) geometry: a layered strip. In both these sections, the arguments rely on the mathematical analysis of the situation. Essentially, parameter-independent bounds lead to subsequences that can identify in the limit the correct homogenized equations. The employed reasoning works only for linear models posed in one-dimensional domains. In the remainder of the chapter, the arguments used are formal in the sense that the asymptotic behavior of the solution is assumed *a priori*. The major advantage is that now we can handle multi-dimensional situations. To start with, we rely on the two-scale asymptotics concepts to homogenize in Section 2.3 a slightly more complex scenario, still related to second-order elliptic equations now posed for a larger class of heterogeneous domains. Unlike the material presented in Section 2.2, the working technique illustrated in Section 2.3 is applicable to a much larger class of partial differential equations, including evolution problems. Sometimes homogenization procedures introduce memory terms. They usually appear in the form of nonlocal integral terms. We discuss such a situation for the case of a linear ordinary differential equation with an oscillating coefficient in Section 2.6.

2.1 Rigorous Homogenization of a One-Dimensional Linearly Coupled System

Let $\Omega = (0,1)$, $f, g \in L^2(\Omega)$, $a \in L^\infty_+(\Omega)$ such that a is 1-periodic and there exist $\alpha, \beta \in (0,\infty)$ with

$$\alpha \leq a(x) \leq \beta \quad \text{for all } x \in \Omega.$$

For any $\varepsilon > 0$, we denote $a(\frac{x}{\varepsilon}) := a_\varepsilon(x)$ for all $x \in \Omega$.

We consider the following setting:

Find the pair $(u_\varepsilon, v_\varepsilon) \in H_0^1(\Omega) \times H_0^1(\Omega)$ satisfying

$$\frac{d}{dx}\left(-a_\varepsilon(x)\frac{du_\varepsilon}{dx}\right) = v_\varepsilon(x) - u_\varepsilon(x) + f(x) \quad \text{in } \Omega, \tag{2.1}$$

$$\frac{d}{dx}\left(-a_\varepsilon(x)\frac{dv_\varepsilon}{dx}\right) = u_\varepsilon(x) - v_\varepsilon(x) + g(x) \quad \text{in } \Omega, \tag{2.2}$$

together with the boundary conditions

$$u_\varepsilon(0) = u_\varepsilon(1) = 0, \tag{2.3}$$

$$v_\varepsilon(0) = v_\varepsilon(1) = 0. \tag{2.4}$$

We refer to this problem as (P_ε), and we often call it "the microscopic problem". This system models the one-dimensional diffusion of two populations of chemical species of similar sizes, say A and B, sharing the spatial domain Ω. For convenience, each population is assumed to move equally fast, i.e. the same diffusion coefficient $a_\varepsilon(\cdot)$ is taken in the structure of the two fluxes. The production by reaction (i.e. the right-hand side of (P_ε)) corresponds to a reversible volumetric chemical reaction mechanism of type $A \rightleftharpoons B$. This situation also arises in the modeling of so-called solid–liquid chemical reactions. Such a right-hand side corresponds then to a production term accounting for deviations from Henry's law or Raoult's law (depending on whether the meaning of the unknowns refers to concentrations or partial pressures); see Section 2.4.3.3 in Muntean (2015) for more modeling details. We do not associate any particular physical meaning to $f(\cdot)$ and $g(\cdot)$. They appear in the model equations as the result of the enforced vanishing of the Dirichlet boundary conditions.

2.1.1 *Well-posedness of the microscopic problem*

Thinking in terms of weak solutions, we prove now that problem (P_ε) is well-posed. Set $w_\varepsilon := u_\varepsilon + v_\varepsilon$. It is useful to note that adding (2.1) and (2.2) and using (2.3) and (2.4), we obtain the following boundary-value problem[1]:

$$\frac{d}{dx}\left(-a_\varepsilon(x)\frac{dw_\varepsilon(x)}{dx}\right) = f(x) + g(x) \quad \text{for } x \in \Omega, \tag{2.5}$$

$$w_\varepsilon(0) = 0 = w_\varepsilon(1). \tag{2.6}$$

A straightforward application of the Lax–Milgram lemma (cf. Theorem 7.1) guarantees the weak solvability of this problem, i.e. there exists a unique $w_\varepsilon \in H_0^1(\Omega)$ satisfying the weak formulation

$$\left(a_\varepsilon(x)\frac{dw_\varepsilon(x)}{dx}, \frac{d\varphi(x)}{dx}\right) = ((f+g)(x), \varphi(x)) \tag{2.7}$$

for all $\varphi \in H_0^1(\Omega)$. Set $p_\varepsilon := u_\varepsilon - v_\varepsilon$. Subtracting now (2.2) from (2.1) leads to the following problem:

$$\frac{d}{dx}\left(-a_\varepsilon(x)\frac{dp_\varepsilon(x)}{dx}\right) + 2p_\varepsilon(x) = f(x) - g(x) \quad \text{for } x \in \Omega, \tag{2.8}$$

$$p_\varepsilon(0) = 0 = p_\varepsilon(1). \tag{2.9}$$

Applied here once more, the Lax–Milgram lemma ensures the existence and uniqueness of a $p_\varepsilon \in H_0^1(\Omega)$ satisfying the weak formulation

$$\left(a_\varepsilon(x)\frac{dp_\varepsilon(x)}{dx}, \frac{d\phi(x)}{dx}\right) + 2(p_\varepsilon(x), \phi(x)) = ((f-g)(x), \phi(x)) \tag{2.10}$$

for all $\phi \in H_0^1(\Omega)$. Consequently, we obtain the unique pair $(u_\varepsilon, v_\varepsilon)$ defined by means of w_ε and p_ε as follows:

$$u_\varepsilon = \frac{1}{2}(w_\varepsilon + p_\varepsilon) \in H_0^1(\Omega),$$

$$v_\varepsilon = \frac{1}{2}(w_\varepsilon - p_\varepsilon) \in H_0^1(\Omega)$$

[1]Note that the homogenization of problems (2.5)–(2.6) is precisely the case discussed in Section 1.3 in Bensoussan *et al.* (1978).

is the desired weak solution to problem (P_ε).[2] Note also that the Lax–Milgram lemma also gives the stability of w_ε and p_ε (and, hence, u_ε and v_ε as well) with respect to the initial data and other parameters in the model.

2.1.2 *Passage to the homogenization limit*

We are interested in what happens with the structure of problem (P_ε) as $\varepsilon \to 0$. We first recall that a_ε is a 1-periodic function. Consequently, we expect that the functions u_ε, v_ε, $\frac{du_\varepsilon}{dx}$, and $\frac{dv_\varepsilon}{dx}$ oscillate as $\varepsilon \to 0$.

The basic questions we ask at this stage are the following:

> Q1: Can we determine the shape of u_ε and v_ε?

> Q2: What happens with the structure of model (P_ε) as $\varepsilon \to 0$? In other words, can we write down what the limit model, say (P_0), looks like (if we can identify a limiting procedure that makes sense)?

> Q3: What happens with the function $a_\varepsilon(\cdot) = a(\frac{\cdot}{\varepsilon})$ as $\varepsilon \to 0$?

Question (Q3) is easier to answer mainly because functional analysis arguments guarantee that $a_\varepsilon \overset{*}{\rightharpoonup} \langle a \rangle$ as $\varepsilon \to 0$; see Lemma 7.2 and its proof. Here, $\langle \cdot \rangle$ is the notation for the following average:

$$\langle \varphi \rangle := \frac{1}{|\Omega|} \int_\Omega \varphi(z)dz \quad \text{for any } \varphi \in L^1(\Omega),$$

where $\Omega \neq \emptyset$ is a bounded measurable subset of $\mathbb{R}^d$, while $|\Omega|$ denotes the d-dimensional Lebesgue measure of the set Ω.

To address questions (Q1) and (Q2), we argue at the level of the equations for w_ε and p_ε, respectively. Let us start with a guessing attempt to address question (Q2). We discuss next the case of w_ε. A somewhat similar discussion can be done for p_ε.

[2]Note that, posed in one-dimensional, (2.5)–(2.6) and (2.8)–(2.9) are exactly solvable, so there is no need *per se* to infer the use of the Lax–Milgram lemma. However, if the same discussion is posed in higher space dimensions, where the set Ω can take a complex shape, the situation becomes more involved and the Lax–Milgram lemma becomes a very useful tool.

Note that Lemma 7.2 may lure us to thinking that maybe as $w_\varepsilon \to w_0$, the corresponding limit (homogenized) problem would be

$$\frac{d}{dx}\left(-\langle a\rangle \frac{dw_0(x)}{dx}\right) = f(x) + g(x) \quad \text{for } x \in \Omega, \tag{2.11}$$

$$w_0(0) = 0 = w_0(1), \tag{2.12}$$

i.e. one replaces the oscillations in the diffusion coefficient with the average. This guess though is not the right one as we see in the following. Imagine for a moment that the equation for w_ε does not involve any differential operator. We would then have

$$-a_\varepsilon w_\varepsilon = f + g.$$

Hence, this gives $w_\varepsilon = -\frac{f+g}{a_\varepsilon}$, which gives, after passing to the limit $\varepsilon \to 0$ via weak convergence, $w_\varepsilon \rightharpoonup -(f+g)\langle \frac{1}{a}\rangle = w_0$. So, we obtain as the limit equation

$$-\frac{1}{\langle \frac{1}{a}\rangle} w_0 = f + g,$$

which is different from

$$-\langle a\rangle w_0 = f + g$$

unless a is the constant function; see Exercise 2.7.3.

The correct answer to (Q2) is given by the following result.

Theorem 2.1. *The sequence $(u_\varepsilon, v_\varepsilon) \in H_0^1(\Omega) \times H_0^1(\Omega)$ satisfying weakly (P_ε) converges in $L^2(\Omega)\times L^2(\Omega)$ to $(u_0, v_0) \in H_0^1(\Omega)\times H_0^1(\Omega)$. The limit pair (u_0, v_0) satisfies weakly the following limit problem referred here as (P_0):*

$$\frac{d}{dx}\left(-\frac{1}{\langle \frac{1}{a}\rangle}\frac{du_0(x)}{dx}\right) = v_0(x) - u_0(x) + f(x) \quad \text{for } x \in \Omega,$$

$$\frac{d}{dx}\left(-\frac{1}{\langle \frac{1}{a}\rangle}\frac{dv_0(x)}{dx}\right) = u_0(x) - v_0(x) + g(x) \quad \text{for } x \in \Omega,$$

$$u_0(0) = 0 = u_0(1), \quad v_0(0) = 0 = v_0(1).$$

Proof. To prove Theorem 2.1, it is convenient to first see that there are constants $c_1, c_2 \in (0, \infty)$, independent of ε, so that

$$||w_\varepsilon||_{H_0^1(\Omega)} \leq c_1 \tag{2.13}$$

and

$$||p_\varepsilon||_{H_0^1(\Omega)} \leq c_2. \tag{2.14}$$

We refer to (2.13) and (2.14) as energy estimates. As a consequence of the Eberlein–Smuljan theorem (see Theorem 7.4), we have the following weak convergences (up to two subsequences):

$$\frac{dw_\varepsilon}{dx} \rightharpoonup \frac{dw_0}{dx} \quad \text{in } L^2(\Omega) \tag{2.15}$$

and

$$\frac{dp_\varepsilon}{dx} \rightharpoonup \frac{dp_0}{dx} \quad \text{in } L^2(\Omega). \tag{2.16}$$

We consider first the equation for w_ε, that is,

$$\frac{d}{dx}\left(-a_\varepsilon(x)\frac{dw_\varepsilon(x)}{dx}\right) = f(x) + g(x) \quad \text{for } x \in \Omega.$$

Fix $x \in \Omega$ arbitrarily. By integrating this equation on the interval $(0, x)$, we obtain

$$-a_\varepsilon(x)\frac{dw_\varepsilon(x)}{dx} = \int_0^x (f(z) + g(z))dz + C_\varepsilon^1.$$

Here, $(C_\varepsilon^1) \subset \mathbb{R}$ is a bounded sequence of real numbers (independent in x but possibly varying with respect to ε). Passing to the limit $\varepsilon \to 0$ in

$$-\frac{dw_\varepsilon(x)}{dx} = \frac{1}{a_\varepsilon(x)}\left[\int_0^x (f(z) + g(z))dz + C_\varepsilon^1\right]$$

with the help of (2.15) leads to

$$-\frac{dw_0(x)}{dx} = \left\langle\frac{1}{a}\right\rangle\left[\int_0^x (f(z) + g(z))dz + C_0^1\right],$$

where $C_0^1 \in \mathbb{R}$ is a limit point of the sequence (C_ε^1). Hence, we get

$$-\frac{1}{\langle\frac{1}{a}\rangle}\frac{dw_0(x)}{dx} = \left[\int_0^x (f(z)+g(z))dz + C_0^1\right].$$

Differentiating the later result with respect to x gives

$$\frac{d}{dx}\left(-\frac{1}{\langle\frac{1}{a}\rangle}\frac{dw_0}{dx}\right) = f+g \quad \text{in } \Omega. \tag{2.17}$$

We now look at the equation for p_ε. From

$$\frac{d}{dx}\left(-a_\varepsilon(x)\frac{dp_\varepsilon(x)}{dx}\right) = -2p_\varepsilon(x)+f(x)-g(x) \quad \text{for } x \in \Omega,$$

we deduce from integration on $(0,x)$ that

$$-a_\varepsilon(x)\frac{dp_\varepsilon(x)}{dx} = -2\int_0^x p_\varepsilon(z)dz + \int_0^x (f(z)-g(z))dz + C_\varepsilon^2,$$

where $(C_\varepsilon^2) \subset \mathbb{R}$ is a bounded sequence of real numbers. As was the case with (C_ε^1), the terms in the sequence (C_ε^2) are independent of x, but they may vary with respect to ε. Dividing the last expression by $a_\varepsilon(x)$ and rearranging the terms gives

$$-\frac{dp_\varepsilon(x)}{dx} + 2\frac{1}{a_\varepsilon(x)}\int_0^x p_\varepsilon(z)dz = \frac{1}{a_\varepsilon(x)}\left[\int_0^x (f(z)-g(z))dz + C_\varepsilon^2\right].$$

Now, we wish to pass in the result to the limit $\varepsilon \to 0$. The main tool here is the weak convergence (2.16). Most of the terms in the previous expression find their corresponding limit, but (2.16) does not provide sufficient information to treat the term $\frac{1}{a_\varepsilon(x)}\int_0^x p_\varepsilon(z)dz$. Using the fact that $p_\varepsilon \to p_0$ strongly in $L^2(\Omega)$ and combining information from Lemmas 7.2 and 7.5, we deduce that

$$\frac{1}{a_\varepsilon}\int_0^x p_\varepsilon(z)dz \rightharpoonup \left\langle\frac{1}{a}\right\rangle \int_0^x p_0(z)dz \quad \text{in } L^2(\Omega). \tag{2.18}$$

We give next a simple direct proof to (2.18). Denote

$$P_\varepsilon(x) := \int_0^x [-2p_\varepsilon(z)]dz,$$

$$P_0(x) := \int_0^x [-2p_0(z)]dz,$$

and take an arbitrary test function, $\varphi \in L^2(\Omega)$. We have the following expression:

$$\begin{aligned}
&\left|\int_0^1 \left(\frac{1}{a_\varepsilon(x)}P_\varepsilon(x) - \left\langle\frac{1}{a}\right\rangle P_0(x)\right)\varphi(x)dx\right| \\
&= \left|\int_0^1 \left(\frac{1}{a_\varepsilon(x)}P_\varepsilon(x) - \frac{1}{a_\varepsilon(x)}P_0(x) + \frac{1}{a_\varepsilon(x)}P_0(x) - \left\langle\frac{1}{a}\right\rangle P_0(x)\right)\varphi(x)dx\right| \\
&\leq \int_0^1 \left|\left(\frac{1}{a_\varepsilon(x)}(P_\varepsilon(x) - P_0(x))\varphi(x)\right)\right| dx \\
&\quad + \left|\int_0^1 \left(\frac{1}{a_\varepsilon(x)} - \left\langle\frac{1}{a}\right\rangle\right)P_0(x)\varphi(x)dx\right|\Bigg) \\
&\leq \left\|\frac{1}{a_\varepsilon}\right\|_{L^\infty(\Omega)} ||P_\varepsilon - P_0||_{L^2(\Omega)}||\varphi||_{L^2(\Omega)} \\
&\quad + \left|\int_0^1 \left(\frac{1}{a_\varepsilon(x)} - \left\langle\frac{1}{a}\right\rangle\right)\Bigg)P_0(x)\varphi(x)dx\right|.
\end{aligned}$$

Correspondingly, the last terms of this last expression vanish[3] as $\varepsilon \to 0$; note that $P_0\varphi \in L^1(\Omega)$.

We can now recover the limit equation

$$\frac{d}{dx}\left(-\frac{1}{\langle\frac{1}{a}\rangle}\frac{dp_0(x)}{dx}\right) = -2p_0(x) + (f(x) - g(x)) \quad \text{for } x \in \Omega. \tag{2.19}$$

Using the trace embedding $H^1(\Omega) \hookrightarrow L^2(\partial\Omega)$, we recover the required boundary conditions for the limit problems:

$$w_0(0) = w_0(1) = p_0(0) = p_0(1) = 0. \tag{2.20}$$

Now, since

$$\begin{aligned}
u_\varepsilon &= \frac{1}{2}(w_\varepsilon + p_\varepsilon) \in H_0^1(\Omega), \\
v_\varepsilon &= \frac{1}{2}(w_\varepsilon - p_\varepsilon) \in H_0^1(\Omega),
\end{aligned}$$

the same weak convergence argument gives

$$\begin{aligned}
u_0 &= \frac{1}{2}(w_0 + p_0) \in H_0^1(\Omega), \\
v_0 &= \frac{1}{2}(w_0 - p_0) \in H_0^1(\Omega).
\end{aligned}$$

---

[3]These arguments can be easily adapted to provide a proof for Lemma 7.5.

This information, combined with (2.17), (2.19), and respectively (2.20), completes the proof of the theorem. □

2.2 Upscaling Gaseous Diffusion through a Two-Dimensional Layered Material

In this section, we present a classical example in the homogenization theory: the upscaling of a linear elliptic equation posed in a layered medium. As physical motivation, we have in mind the upscaling of gaseous diffusion through a periodic two-dimensional layered material, made by alternating strips with two materials of different physical properties.

The presentation of this section follows closely Le Bris (2015). More details on this particular example can be found in Section 5.4 in Cioranescu and Donato (1999).

2.2.1 *Motivation and main result*

Take $\delta > 0$ arbitrarily fixed. Let $\Omega^\delta := (0,1) \times (0,\delta)$ be a domain with a layered structure. Such a domain is also referred to as lamellar or laminated. It is composed of alternating layers of two different materials. The geometry of the composite we have in mind is depicted in Figure 2.1. The basic assumptions behind this geometry are twofold:

(i) Each material has homogeneous (invariant) properties along the Ox_2 axis.
(ii) The composite is $\varepsilon = \frac{\ell}{L}$-periodic along the Ox_1 axis. Here, Ox_1 and Ox_2 are the coordinate axes used in Figure 2.1.

Our microscopic problem reads as follows:

Find $u_\varepsilon \in H_0^1(\Omega^\delta)$ satisfying the following boundary-value problem:

$$\operatorname{div}(-a_\varepsilon(x)\nabla u_\varepsilon(x)) = f(x) \quad \text{for } x \in \Omega^\delta, \tag{2.21}$$

$$u_\varepsilon(x) = 0 \quad \text{for } x \in \partial\Omega^\delta, \tag{2.22}$$

where, for $x = (x_1, x_2)$, we have

$$a_\varepsilon(x) := \begin{pmatrix} a\left(\frac{x_1}{\varepsilon}\right) & 0 \\ 0 & a\left(\frac{x_1}{\varepsilon}\right) \end{pmatrix}. \tag{2.23}$$

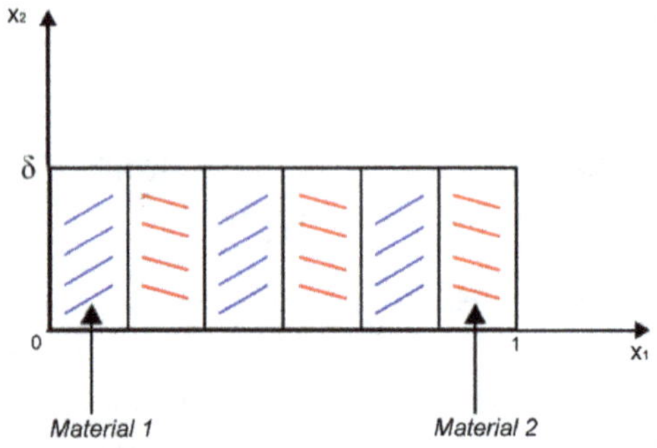

Figure 2.1. Sketch of the layered domain Ω^δ, made by replicating a two-strip pattern.

It is worthwhile to note that the entries in the matrix a_ε are independent of the variable x_2. Also, the partial differential equation admits the rewriting

$$\operatorname{div}\left(-a_\varepsilon(x)\left(\frac{\partial u_\varepsilon}{\partial x_1}e_1+\frac{\partial u_\varepsilon}{\partial x_2}e_2\right)\right)=f,$$

where $\{e_1, e_2\}$ is a system of orthonormal vectors in $\mathbb{R}^2$.

We work with the following hypotheses:

- $0 < c_1 \leq a(x_1) \leq c_2 < \infty$ for all $x_1 \in [0, 1]$;
- $a(\cdot)$ is a 1-periodic function with $a(x_1) = \begin{cases} \alpha, & x_1 \in [0, \frac{1}{2}] \\ \beta, & x_1 \in (\frac{1}{2}, 1] \end{cases}$;
- $f \in L^2(\Omega^\delta)$ (or, alternatively, $f \in H^{-1}(\Omega^\delta)$).

Problem (2.21) has the following physical meaning: u_ε is, for instance, the mass concentration of a population of $O_2(g)$ particles diffusing through a layered packaging material Ω^δ, a_ε is the diffusion coefficient for a pair of two consecutive strips, while f is a uniform (in ε) volumetric production term for $O_2(g)$. Depending on the case, it can be either a sink (trapping) or a source.

We ask ourselves the following:

> What happens with problem (2.21), as well as its solution u_ε, as $\varepsilon \to 0$?

Theorem 2.2. *The weak solution $u_\varepsilon \in H_0^1(\Omega^\delta)$ to problem (2.21) converges to the weak solution of the following problem:*

Find $u_0 \in H_0^1(\Omega^\delta)$ *such that*

$$\operatorname{div}(-\mathbb{D}(x)\nabla u_0(x)) = f(x) \quad \text{in } \Omega^\delta, \tag{2.24}$$

$$u_0 = 0 \quad \text{at } \partial\Omega^\delta, \tag{2.25}$$

where

$$\mathbb{D} = \begin{pmatrix} \frac{1}{\langle a \rangle} & 0 \\ 0 & \langle a \rangle \end{pmatrix}. \tag{2.26}$$

Equation (2.24) is the upscaled or homogenized problem, while the matrix $\mathbb{D}$ is referred to as the effective diffusion coefficient. They are direct results of the passage to the homogenization limit, i.e. letting ε go to zero. The fact that the detailed structure of $\mathbb{D}$ becomes known is not only beautiful but a quite useful information for practical purposes, turning the periodic homogenization technique a useful tool for engineers. It is worth insisting on the fact that this information is accessible because of the assumption $a_\varepsilon(x) = a\left(\frac{x_1}{\varepsilon}\right)$. In the case that a_ε depends on both variables in an arbitrary way so that the entries are essentially bounded and the matrix is both positive definite and coercive, then one can still show (via some sort of G-convergence reasoning) the existence of a limit coefficient $\mathbb{D}$, but no information is obtainable about the entries of this effective matrix; see Proposition 2.1.2 in Le Bris (2015). In Section 2.3, we will see that if instead of the regular arrangement of strips we take a periodic arrangements of inclusions or holes (perforations), one can still obtain detailed information on the structure of the effective coefficient. Observe that if (2.24) is replaced by the linear elasticity equations as they are applied to a laminated composite, then the entries of $\mathbb{D}$ can be identified precisely as for an orthotropic material; see e.g. Section 13 of Chapter 3 from Chechkin *et al.* (2007).

2.2.2 *Proof of Theorem 2.2*

In this section, we provide the details of the proof of Theorem 2.2.

Proof. Testing (2.21) with $\varphi \in H_0^1(\Omega^\delta)$ gives

$$\int_{\Omega^\delta} a_\varepsilon \nabla u_\varepsilon \nabla \varphi dx = \int_{\Omega^\delta} f \varphi dx.$$

We prove now an energy estimate by choosing in the previous identity as test function $\varphi = u_\varepsilon$. It yields

$$\begin{aligned} c_1||\nabla u_\varepsilon||^2_{L^2(\Omega^\delta)} &\leq ||f||_{L^2(\Omega^\delta)}||u_\varepsilon||_{L^2(\Omega^\delta)} \\ &\leq c_P||f||_{L^2(\Omega^\delta)}||\nabla u_\varepsilon||_{L^2(\Omega^\delta)}, \end{aligned}$$

which ensures

$$||\nabla u_\varepsilon||_{L^2(\Omega^\delta)} \leq \frac{c_P}{c_1}||f||_{L^2(\Omega^\delta)}. \tag{2.27}$$

The upper bound on (2.27) is independent of the choice of the parameter ε. Here, c_P is the constant in Poincaré's inequality. This constant depends on $\sqrt{\delta}$, while it is independent of ε. Since u_ε is uniformly bounded with respect to ε in $L^2(\Omega^\delta)$ and ∇u_ε is also uniformly bounded with respect to ε in $L^2(\Omega^\delta)$, the Rellich–Kondrachov result (cf. Theorem 7.5) ensures that, up to subsequences, $u_\varepsilon \to u_0$ in $L^2(\Omega^\delta)$ and $\nabla u_\varepsilon \rightharpoonup \nabla u_0$ in $L^2(\Omega^\delta)$ as $\varepsilon \to 0$. In this way, the uniform bound (2.27) gives hope that the homogenization process is, in principle, possible.

We denote the components of the flux in (2.21) by

$$\sigma_i^\varepsilon(x_1, x_2) = -a\left(\frac{x_1}{\varepsilon}\right)\frac{\partial}{\partial x_i}u_\varepsilon(x_1, x_2),$$

with $i \in \{1, 2\}$. To understand where σ_i^ε is converging as $\varepsilon \to 0$, we need to handle the convergence of the product of two oscillating sequences $(a\left(\frac{x_1}{\varepsilon}\right))_\varepsilon$ and $\left(\frac{\partial u_\varepsilon}{\partial x_i}\right)_\varepsilon$. As neither of these sequences converges strongly to the limit, to elucidate what the limit point is, we need to rely on a couple of subtle arguments that essentially use the structure of the microscopic problem.

We first observe that

$$-\frac{1}{a\left(\frac{x_1}{\varepsilon}\right)}\sigma_1^\varepsilon = \frac{\partial u_\varepsilon}{\partial x_1} \rightharpoonup \frac{\partial u_0}{\partial x_1}.$$

On the other hand, we have $a \in L^\infty(\Omega^\delta)$, with $a \geq c_1 > 0$ and $\frac{\partial u_\varepsilon}{\partial x_1} \in L^2(\Omega^\delta)$. It also holds that $||\sigma_1^\varepsilon||_{L^2(\Omega^\delta)} \leq c$. From

$$\frac{\partial \sigma_1^\varepsilon}{\partial x_1} = f - \frac{\partial \sigma_2^\varepsilon}{\partial x_2},$$

we obtain

$$f - \frac{\partial \sigma_2^\varepsilon}{\partial x_2} \in L^2_{x_1}(H^{-1}_{x_2}). \tag{2.28}$$

Finally, from

$$\sigma_1^\varepsilon \in L^2_{x_1} \otimes L^2_{x_2} = L^2_{x_1}(L^2_{x_2}),$$
$$\frac{\partial \sigma_1^\varepsilon}{\partial x_1} \in L^2_{x_1} \otimes H^{-1}_{x_2} = L^2_{x_1}(H^{-1}_{x_2}),$$

we deduce that, at least up to a subsequence, the following convergence holds:

$$\sigma_1^\varepsilon \to \sigma_1 \text{ strongly in } L^2_{x_1}(H^{-1}_{x_2}). \tag{2.29}$$

As a consequence of this strong convergence and recalling additionally that $a\left(\frac{x_1}{\varepsilon}\right) \rightharpoonup \left\langle \frac{1}{a} \right\rangle$, we obtain

$$a\left(\frac{x_1}{\varepsilon}\right)\sigma_1^\varepsilon \rightharpoonup \left\langle \frac{1}{a} \right\rangle \sigma_1. \tag{2.30}$$

Summarizing these arguments, based on the facts $-\frac{1}{a\left(\frac{x_1}{\varepsilon}\right)}\sigma_1^\varepsilon = \frac{\partial u_\varepsilon}{\partial x_1} \rightharpoonup \frac{\partial u_0}{\partial x_1}$ and $a\left(\frac{x_1}{\varepsilon}\right)\sigma_1^\varepsilon \rightharpoonup \left\langle \frac{1}{a} \right\rangle \sigma_1$, we can finally identify the first component of the flux σ_1 in the upscaled equation as

$$\sigma_1 = -\frac{1}{\left\langle \frac{1}{a} \right\rangle} \frac{\partial u_0}{\partial x_1}.$$

To identify the structure of the limit flux σ_2, we proceed as follows. Accounting for the independence of a on x_2, we write

$$\sigma_2^\varepsilon = -a\left(\frac{x_1}{\varepsilon}\right)\frac{\partial}{\partial x_2} u_\varepsilon(x_1, x_2).$$

Since $u_\varepsilon \to u_0$ in $L^2(\Omega^\delta)$ and $a\left(\frac{x_1}{\varepsilon}\right) \overset{*}{\rightharpoonup} \langle a \rangle$ in $L^\infty(\Omega^\delta)$, it also holds that $a\left(\frac{x_1}{\varepsilon}\right) u_\varepsilon \rightharpoonup \langle a \rangle u_0$ in $L^2(\Omega^\delta)$. Hence,

$$\sigma_2^\varepsilon \rightharpoonup \sigma_2 = -\langle a \rangle \frac{\partial u_0}{\partial x_2} \quad \text{in } H^{-1}(\Omega^\delta),$$

which completes the proof. □

2.2.3 *A tale of two other limits: $\delta \to 0$ and $\delta \to \infty$*

A natural continuation of the discussion concerning the diffusion through the layered geometry from Section 2.2 would be to study the dimension reduction question, i.e. letting $\delta \to 0$, for either the microscopic problem or the upscaled problem, or for both of them. This is a singular perturbation limiting case. Proving δ-independent bounds as done by van Duijn and Pop (2004) and Fatima *et al.* (2014) can clarify the structure of the reduced limit problem; see also Chapter 5 for a related dimension reduction exercise. The situation becomes much more difficult if we would allow $\delta = \delta(\varepsilon)$ and then aim to perform the simultaneous homogenization and dimension-reduction limit; a related work proceeding in a similar direction is that by Gustafsson and Mossino (2003).

Another potentially interesting limiting case is $\delta \to \infty$. Although this connects information between bounded and unbounded domains, it is rather a regular perturbation-type limiting case. To handle this limit in a rigorous fashion, we refer the reader for instance to the well-written textbook by Chipot (2002), which contains the working techniques needed to completely clarify the situation.

2.3 Formal Asymptotic Homogenization, or How to Guess the Averaged Model: A Direct Approach

In this section, we use asymptotic expansions to perform the homogenization process for partial differential equations. Such asymptotic expansions typically involve two separate spatial scales: one *micro* and one *macro*. They can be modified to account for further intermediate scales as well. A careful study of Section 2.3 will give the reader a good grip on the methodology. If one believes that solutions to partial differential equations eventually admit representations that can be written in terms of two-scale expansions, then these expansions simply "dislocate" the inherently involved oscillations and successfully unveil the structure of upscaled model equations and effective coefficients. The basic working assumption is the periodicity of the coefficients and/or of the arrangement of the array of microstructures. We propose in this section a formal way of working; mathematical analysis work is required to justify the validity of the two-scale expansion so that trust in the obtained results can be gained. This will be done in Chapters 3 and 4. The content of this section[4] largely follows Muntean and Chalupecky (2011) and is also inspired very much by the excellently written texts by

[4] It is worth noting that Section 2.3 can be read independently of the other parts of this textbook. The material presented here can be taught to non-mathematicians, the only prerequisite being a good command of basic calculus.

Neuss-Radu (1992), Hornung (1997), and Pavliotis and Stuart (2008). For a supplementary reading material, see Persson *et al.* (1993).

2.3.1 *A model problem: Stationary diffusion and chemical reactions in perforated media*

We introduce a formal asymptotics technique, which is sometimes called "two-scale asymptotic homogenization". Starting off from microscopic models whose validity we trust, we can derive using this technique the structure of macroscopic models, as well as constitutive laws for their effective coefficients. It is worth noting that the applicability of the asymptotic expansions method is rather generous. The approach is highly successful for a wide class of second-order partial differential equations, ranging, for instance, from the elasticity equations to free-boundary problems modeling phase transitions in heterogeneous materials. The methodology has been applied to higher-order partial differential equations as well, sometimes oscillations being allowed in both space and time variables, cf. for instance Holmbom (1997). The cases of ordinary differential equations and first-order partial differential equations are slightly peculiar, as among them there are many examples of non-homogenizable equations.

Instead of presenting the technique in the context of a complex model, we prefer to apply it to a simpler example, though remaining relevant to the topics connected to reactive transport and storage in porous media. Concretely, we consider a model problem that describes the interplay between the stationary diffusion of a population of gaseous molecules (called u_ε) and a combination of surface and volume productions through chemical reactions where these molecules become involved.

Denoting the porous media by $\Omega_\varepsilon \subset \mathbb{R}^d$, our toy problem reads as follows:

Find u_ε such that the following equations are satisfied:

$$\operatorname{div}(-a_\varepsilon \nabla u_\varepsilon) = f_\varepsilon \quad \text{in } \Omega_\varepsilon, \tag{2.31}$$

together with the boundary conditions

$$-a_\varepsilon \nabla u_\varepsilon \cdot n_\varepsilon = g_\varepsilon \quad \text{on } \Gamma_\varepsilon, \tag{2.32}$$

$$u_\varepsilon = 0 \quad \text{on } \Gamma^{ext}. \tag{2.33}$$

The sets $\Omega_\varepsilon, \Gamma_\varepsilon$, and Γ^{ext} will be defined[5] in Section 2.3.2. The coefficient a_ε refers to the diffusivity of the gaseous molecules, while the terms f_ε and g_ε denote productions by volume and surface chemical reaction, respectively.

The aim of the following section is to use the concept of perforated domains to approximate the internal structure of a regular heterogeneous porous media. Such a perforated domain is the place where these model equations are posed. To fix ideas, we look at the simplest porous media: Imagine a composite material made of an homogeneous background phase (say air or water) and an array of periodically placed inclusions (voids, impermeable grains, etc.).

2.3.2 *Perforated domains*

We describe the geometry of a perforated domain $\Omega_\varepsilon \subset \mathbb{R}^d$. Two examples of two-dimensional sets Ω_ε are indicated in Figure 2.2. The right panel fits best to our purpose here.

Let $Z \subset \mathbb{R}^d$ be a hypercube. Furthermore, let $k := (k_1, \ldots, k_d) \in \mathbb{Z}^d$ be a vector of indices and $e := (e_1, \ldots, e_d)$ be the set of independent orthonormal unit vectors in $\mathbb{R}^d$. For a given subset $X \subset Z$, we denote by X^k the shifted subset

$$X^k := X + \sum_{i=1}^{d} k_i e_i.$$

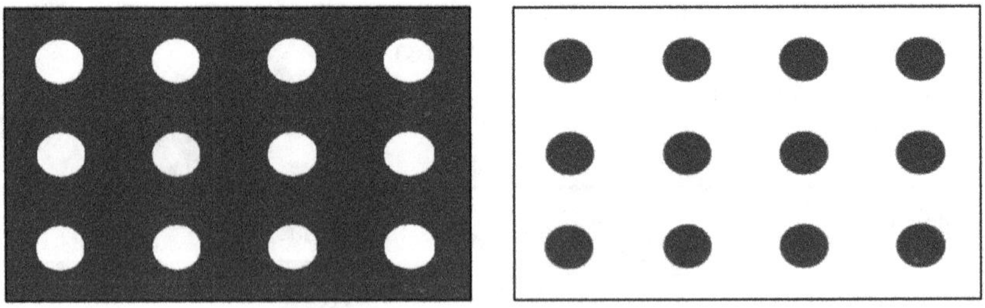

Figure 2.2. Sketch of two perforated structures Ω_ε; each color indicates a different material. Think of the white part being filled with air and the black part being made of a solid material.

[5]We discuss the geometry of the perforated domain in various places in this textbook, all agreeing with the content of Section 7.1. Methodologically, this seems to be a critical part when teaching homogenization-based techniques.

Using the writing in terms of shifted subsets, the pore matrix (or pore skeleton) of the target porous media can be represented by

$$\Omega_0^\varepsilon := \bigcup_{k\in\mathbb{Z}^d} \{\varepsilon Y_0^k : Y_0^k \subset \Omega\}.$$

Consequently, the total pore space can be represented by

$$\Omega_\varepsilon := \Omega \backslash \Omega_0^\varepsilon.$$

The set εY_0^k is sometimes referred to in the literature as the ε-homotetic set of Y_0^k. The total (inner) surface of the skeleton is denoted by

$$\Gamma^\varepsilon := \partial\Omega_0^\varepsilon = \bigcup_{k\in\mathbb{Z}^d} \{\varepsilon\Gamma^k : \varepsilon\Gamma^k \subset \Omega\},$$

where the object Γ^k is the shifted interface

$$\Gamma^k := \Gamma + \sum_{i=1}^{d} k_i e_i.$$

Γ^ε accounts for the total amount of interface available for the (surface) chemical reaction to occur. We introduce the unit normal vector $n_\varepsilon = n\left(\frac{x}{\varepsilon}\right)$ corresponding to Γ^ε. Furthermore, let Γ^{ext} denote the outer boundary of Ω. As a side remark, note that we often use both notations Γ^{ext} and $\partial\Omega$ to indicate the same boundary.

Now, it is a good moment for the reader to imagine what would be the effect of varying ε on the structure of Ω_ε. Essentially, when ε decreases, the number of inclusions (suitably scaled with ε) present in the background domain Ω increases, while keeping the volumetric porosity fixed. This makes the transition of the mathematical representation of the porous domain from a heterogeneous region to a homogeneous region; see the chessboards example shown in Figure 1.4.

2.3.3 *Microscopic model equations*

We begin with the following formulation: Find

$$u_\varepsilon \in H^1(\Omega_\varepsilon; \Gamma^{ext}) := \{v \in H^1(\Omega_\varepsilon) : v = 0 \quad \text{at } \Gamma^{ext}\}$$

such that the following linear elliptic equation is satisfied weakly:

$$\operatorname{div}(-a_\varepsilon \nabla u_\varepsilon) = f_\varepsilon \quad \text{in } \Omega_\varepsilon, \tag{2.34}$$

together with the boundary conditions

$$-a_\varepsilon \nabla u_\varepsilon \cdot n_\varepsilon = g_\varepsilon \quad \text{on } \Gamma_\varepsilon, \tag{2.35}$$

$$u_\varepsilon = 0 \quad \text{on } \Gamma^{ext}. \tag{2.36}$$

We take $a_\varepsilon \in L^\infty_+(\Omega_\varepsilon)$, $f_\varepsilon \in L^2(\Omega_\varepsilon)$, $g_\varepsilon \in L^2(\Gamma_\varepsilon)$, all with uniformly bounded corresponding norms, and refer to (2.34)–(2.36) as problem (P_ε).

Note that $\partial\Omega$ is a hypersurface in $\mathbb{R}^d$ and $\lambda^{d-1}(\Gamma_\varepsilon) \neq 0$, with λ^{d-1} being the $d-1$-dimensional Lebesgue measure. Furthermore, $\partial\Omega_\varepsilon = \Gamma^{ext} \cup \Gamma_\varepsilon$, where $\Gamma^{ext} \cap \Gamma_\varepsilon = \emptyset$. We take

$$f_\varepsilon(x) := f_0(x) + \varepsilon f_1\left(x, \frac{x}{\varepsilon}\right),$$

$$g_\varepsilon(x) := \varepsilon g_0\left(x, \frac{x}{\varepsilon}\right),$$

and we assume some smoothness for $f_1(\cdot,\cdot)$ and $g_0(\cdot,\cdot)$, particularly with respect to their second variable.

We consider[6] that the geometry of the porous medium, model parameters, and boundary data are prescribed such that our microscopic problem (P_ε) is well-posed in $H^1(\Omega_\varepsilon;\Gamma^{ext})$. In this context, the term f_ε represents a volumetric chemical reaction acting in the bulk of the available domain, while the term g_ε points to a surface chemical reaction. For the simplicity of the exposure, we assume that the functions f_ε and g_ε depend only on the x-variable. From a formal asymptotics point of view, only minimal modifications are needed to adapt the following calculations for the case when f_ε and g_ε also depend on both t and u_ε. Additionally, we take the functions f_0, f_1, and g_0 to be of the order of $\mathcal{O}(1)$. It is important to note that the boundary production has the structure $\varepsilon^\alpha g_0(\cdot,\cdot)$, with $\alpha = 1$. This is the only scaling in the small parameter ε that leads to an interesting limit. To support informally this statement, we can easily see that

$$\lim_{\varepsilon\to 0} \varepsilon\lambda^{d-1}(\Gamma_\varepsilon) = \lambda^{d-1}(\Gamma)\frac{\lambda^d(\Omega)}{\lambda^d(\Gamma)},$$

where λ^m is the m-dimensional Hausdorff measure. The homogenization limit assumes that as ε vanishes, the volume of the initial region Ω remains

[6] Note that some more information must be provided, e.g. on $a_\varepsilon(\cdot)$, to be able to say anything about the desired well-posedness. Such matters will be discussed in detail in Chapter 3, where the passage to the homogenization limit is carried out rigorously.

unchanged, while the number of inclusions (holes) increases gradually. Consequently, the amount of interior surface available for interfacial chemical reactions to occur increases. If $\alpha \in (-\infty, 1)$, we expect that such a boundary term will force the solution to blow up as ε reaches 0. On the other hand, if $\alpha \in (1, +\infty)$, then the boundary term vanishes in the homogenization limit. We will see in Section 2.3.4 that the case of $\alpha = 1$ leads to a non-vanishing limiting term; therefore, this is an interesting setting to explore further.

We refer to (2.34)–(2.36) as problem (P_ε). The limit problem obtained as $\varepsilon \to 0$ is denoted correspondingly by (P_0).

2.3.4 *The asymptotics homogenization procedure: How to guess the structure of macroscopic models?*

The task is now to perform the homogenization process, i.e. we let $\varepsilon \to 0$ in (P_ε). The key idea is to rely on the following *ansatz* (often called the *homogenization ansatz*): Assume that u_ε has the structure

$$u_\varepsilon(x) = u_0(x,y) + \varepsilon u_1(x,y) + \varepsilon^2 u_2(x,y) + \mathcal{O}(\varepsilon^3)\Big|_{\text{for } y:=\frac{x}{\varepsilon}}, \tag{2.37}$$

where, for all $\ell \in \mathbb{Z}$, the modes $u_\ell(x,\cdot)$ of the homogenization expansion are ε-periodic in the second variable and are all of the order of $\mathcal{O}(1)$. Expression (2.37) reminds us remotely of the Fourier expansion of a periodic function with respect to the trigonometric orthonormal system; the more modes $u_\ell(x,\cdot)$ this expression contains, the better the oscillations of u_ε are represented.

Note that if ψ is a sufficiently smooth function (e.g. $\psi \in C^1(\Omega_\varepsilon) \cap C^0(\overline{\Omega}_\varepsilon)$), then the following calculation rule applies:

$$\frac{d}{dx}\psi\left(x,\frac{x}{\varepsilon}\right) = \partial_x\psi\left(x,\frac{x}{\varepsilon}\right) + \frac{1}{\varepsilon}\partial_y\psi\left(x,\frac{x}{\varepsilon}\right)\Big|_{\text{for } y:=\frac{x}{\varepsilon}}. \tag{2.38}$$

To simplify the writing, we will not carry with us the expression "where $y = \frac{x}{\varepsilon}$", but we will implicitly mean it. Essentially, we imagine x and y to be independent variables.[7]

[7]Note that x and y become independent only when the limit $\varepsilon = 0$ is reached. In this limiting case, the perforated domain Ω_ε unfolds into the domain $\Omega \times Y$.

On inserting *ansatz* (2.37) into (P_ε), we get for (2.34) the following structure posed in Ω_ε:

$$\left[\frac{1}{\varepsilon^2}T_{-2} + \frac{1}{\varepsilon}T_{-1} + \varepsilon^0 T_0\right](u_0, u_1, u_2, \ldots) = [f_0 + \varepsilon f_1](u_0, u_1, u_2, \ldots). \tag{2.39}$$

Here, T_{-2}, T_{-1}, and T_0 are shorthand notations for three auxiliary problems that we will define in this section. We obtain

$$\left(\nabla_x + \frac{1}{\varepsilon}\nabla_y\right)\cdot\left(A(y)\left(\nabla_x + \frac{1}{\varepsilon}\nabla_y\right)(u_0 + \varepsilon u_1 + \varepsilon^2 u_2 + \text{h.o.t.})\right)$$
$$= f_0 + \varepsilon f_1 + \mathcal{O}(\varepsilon^2).$$

Consequently, this yields

$$\left(\nabla_x + \frac{1}{\varepsilon}\nabla_y\right)\cdot\left(\frac{1}{\varepsilon}A(y)\nabla_y u_0 + \varepsilon^0\big(A(y)\nabla_x u_0 + A(y)\nabla_y u_1\big)\right.$$
$$\left. + \varepsilon^1\big(A(y)\nabla_x u_1 + A(y)\nabla_y u_2\big) + \text{h.o.t}\right)$$
$$= f_0 + \text{h.o.t.}$$

As the next step, we group the terms by collecting those having the same powers of ε and get that

$$\frac{1}{\varepsilon^2}\nabla_y\cdot\big(A(y)\nabla_y u_0\big)$$
$$+ \frac{1}{\varepsilon}\Big(\nabla_y\cdot\big(A(y)\nabla_x u_0 + A(y)\nabla_y u_1\big) + \nabla_x\cdot\big(A(y)\nabla_y u_0\big)\Big)$$
$$+ \varepsilon^0\Big(\nabla_x\cdot\big(A(y)\nabla_x u_0 + A(y)\nabla_y u_1\big) + \nabla_y\cdot\big(A(y)\nabla_x u_1 + A(y)\nabla_y u_2\big)\Big)$$
$$= f_0 + \text{h.o.t.}$$

We proceed in the same way with the boundary conditions.[8] We obtain

$$-A(y)\left(\nabla_x + \frac{1}{\varepsilon}\nabla_y\right)(u_0 + \varepsilon u_1 + \varepsilon^2 u_2) + \text{h.o.t.} = \varepsilon g_0 + \varepsilon^2 g_1,$$

[8]A hint to speed up calculations: If the microscopic problem is posed in divergence form and has flux boundary conditions, then one can apply the homogenization *ansatz* first to the boundary conditions and then use the result to deal with the partial differential equation.

which leads to the following relations holding at ∂Y_0:

$$\begin{aligned}
\frac{1}{\varepsilon}: &\quad -A(y)\nabla_y u_0 \cdot n(y) = 0,\\
\varepsilon^0: &\quad -A(y)\nabla_x u_0 \cdot n(y) - A(y)\nabla_y u_1 \cdot n(y) = 0,\\
\varepsilon^1: &\quad -\big(A(y)\nabla_x u_1 + A(y)\nabla_y u_2\big)n(y) = g_0(x, y).
\end{aligned}$$

We now need to evaluate the limits of the indeterminate form (2.39). This is only possible in the limit of $\varepsilon \to 0$. A necessary condition for this to happen is that the auxiliary problems

$$\begin{aligned}
T_{-2}(u_0) &= 0,\\
T_{-1}(u_0, u_1) &= 0, \qquad (2.40)\\
T_0(u_0, u_1, u_2) &= F_0
\end{aligned}$$

hold true in $\Omega \times Y$. Here, Ω is the initial domain without the perforations, Y is the microstructure within the periodic cell Z, and F_0 is a zero-order term in ε, including f_0 and eventually some boundary terms. The precise structure of F_0 will become clear toward the end of this averaging procedure. All auxiliary problems are endowed with periodic boundary conditions across ∂Y, and they all depend[9] on the parameter $x \in \Omega$. Having a glance at (2.40), we expect already a certain structure for the macroscale equation. However, at this stage, this structure is too complicated. We will see that the auxiliary problems help in devising the structure of the homogenized equation and will lead to explicit expressions of the effective coefficients.

We examine the first auxiliary problem, i.e. we look at the problem corresponding to T_{-2}:

Find the $u_0 = u_0(x, y)$ solution of

$$\nabla_y \cdot \big(A(y)\nabla_y u_0(x, y)\big) = 0 \quad \text{for } y \in Y, \qquad (2.41)$$

$$-A(y)\nabla_y u_0(x, y) \cdot n(y) = 0 \quad \text{for } y \in \partial Y_0, \qquad (2.42)$$

$$u_0(x, \cdot) \text{ is } Y\text{-periodic for each given } x \in \Omega. \qquad (2.43)$$

To ensure the uniqueness of solutions to (2.41)–(2.43), we impose

$$\int_Y u_0(x, y)dy = c, \qquad (2.44)$$

[9]This dependence keeps at least the measurability property with respect to the x-variable.

where $c \in \mathbb{R}$ is a constant. The structure of (2.41)–(2.43) indicates that u_0 is independent of the choice of y, and hence, we expect the function

$$u_0 = u_0(x) \tag{2.45}$$

to be the macroscopic solution we are looking for. Consequently, the constant c arising in (2.44) becomes a function of x, that is, $c(x) := u_0(x)|Y|$. However, this fact does not have any further practical implications.

To see that (2.45) indeed holds true, we proceed as follows. We multiply (2.41) by $u_0(x, y)$ and integrate the result on Y. Benefiting from (2.41)–(2.43), the integration by parts of the result gives for an $\alpha > 0$ (ellipticity constant of the matrix $A(\cdot)$) the following inequalities:

$$0 = \int_Y A(y)\nabla_y u_0(x, y)\nabla_y u_0(x, y)dy \geq \alpha \int_Y |\nabla_y u_0(x, y)|^2 dy \geq 0.$$

This argument leads to the desired result.

We examine now the next auxiliary problem, which is the one corresponding to T_{-1}:

Find the $u_1 = u_1(x, y)$ solution of

$$\nabla_y \cdot \big(- A(y)\nabla_y u_1\big) = \nabla_y \cdot \big(A(y)\nabla_x u_0\big), \tag{2.46}$$

$$-A(y)\nabla_y u_1(x, y) \cdot n(y) = A(y)\nabla_x u_0(x) \cdot n(y), \quad \text{at } \partial Y_0, \tag{2.47}$$

$$u_1(x, \cdot) \;\text{ is } Y\text{-periodic for each given } x \in \Omega. \tag{2.48}$$

At this point, we recall a theoretical result that ensures the existence and uniqueness of weak solutions to linear elliptic boundary-value problems with periodic boundary conditions, namely Lemma 7.1. This intimately connects to the so-called Fredholm's alternative. We refer the reader to Section 7.3 for more context.

We apply now Lemma 7.1 to problem T_{-1}. Rearranging the terms in (2.46), we obtain after applying Gauß's theorem that

$$-\int_{\partial Y} A(y)\nabla_y u_1 \cdot n(y)d\sigma_y = \int_{\partial Y_0} A(y)\nabla_x u_0 \cdot n(y)d\sigma_y, \tag{2.49}$$

where $d\sigma_y$ is the surface measure on ∂Y. Hence, Lemma 7.1 ensures the existence of a unique (weak) solution u_1 to T_{-1}.

To handle this new situation, it is convenient to introduce the so-called *cell functions*, which are solutions to *cell problems*. Namely, we can rewrite[10] u_1 as

$$u_1(x, y) = W(y) \cdot \nabla_x u_0(x) + \tilde{u}_1(x), \tag{2.50}$$

with $\tilde{u}_1(\cdot)$ being some given function. Here, $W(y)$ is usually referred to as *cell function*

$$W(y) = \big(W_1(y), \ldots, W_d(y)\big)^t \in \mathbb{R}^d.$$

We now introduce a key ingredient – the *cell problem.*

For all $j \in \{1, \ldots, d\}$, find the cell functions $W_j \in H^1_{\#}(Y)$ such that the cell problem

$$\nabla_y \cdot \Big(- A(y)\nabla_y W_j(y)\Big) = \nabla_y \cdot \Big(A(y)e_j\Big) \quad \text{for } y \in Y, \tag{2.51}$$

$$-A(y)\nabla_y W_j(y) \cdot n(y) = A(y)e_j \cdot n(y) \quad \text{for } y \in \partial Y_0, \tag{2.52}$$

$$W_j \text{ is } Y\text{-periodic} \quad \int_Y W_j(y)dy = 0, \tag{2.53}$$

is satisfied weakly.

Observe that the weak solution to the cell problem stated here can be computed independently of the knowledge of the microscopic problem. $W(\cdot)$ only depends on the geometry of the microscopic cell Y, as well as on the coefficient $A(\cdot)$. We can rewrite the cell problem in a slightly more compact form as follows:

---

[10]You can deduce the structure of u_1 easily if you recall the way one obtains general analytic solutions to non-homogeneous ordinary differential equations using the structure of the solution to the corresponding homogeneous problem. We invite the reader to replace the proposed structure for u_1 (2.50) with T_{-1} and deduce explicitly the structure of the cell problem.

Find $W \in H^1_{\#}(Y)$ such that it holds

$$\nabla_y \cdot \Big[- A(y)\Big(\nabla_y W(y) + \mathbb{I}\Big)\Big] = 0 \quad \text{for } y \in Y, \tag{2.54}$$

$$-A(y)\Big(\nabla_y W(y) + \mathbb{I}\Big) \cdot n(y) = 0 \quad \text{for } y \in \partial Y_0, \tag{2.55}$$

$$W \text{ is } Y\text{-periodic} \quad \int_Y W(y)dy = 0, \tag{2.56}$$

We will discuss in Section 2.4 further details concerning the cell problem.

Regarding the particular homogenization example discussed in this section, u_1 appears exclusively under the differentiation with respect to y. Hence, the function $\tilde{u}_1(x)$ arising in (2.50) can be taken to be zero. Note though that when dealing with non-periodic homogenization scenarios, this function can no longer be neglected, as it will naturally depend on the variable y as well.

Considering the partial differential equation for u_2, namely in the auxiliary problem T_0, we note that the following solvability condition has to be fulfilled:

$$\int_Y [\text{r. h. s.}]dy = \int_{\partial Y} [\text{bdd. term}]d\sigma_y. \tag{2.57}$$

After a careful glance, we should recognize that (2.57) is actually a macroscopic (upscaled) equation in the sense that the microstructure variable $y \in Y$ is averaged out from the equation.

It is worth noting that the structure of the macroscopic equation (e.g. (2.57)) can be obtained directly by integrating over the set Y the equation for u_2, that is, the coefficient of the order of $\mathcal{O}(\varepsilon^0)$ in the expanded microscopic problem.

2.3.5 *Structure of the upscaled model equations: Explicit formulas for the effective coefficients*

It remains to derive the precise structure of the macroscopic equation and the associated boundary conditions. To do so, we need to handle the partial differential equation for u_2. In other words, we need to make use of the auxiliary problem T_0, which reads as follows:

Find the $u_2 = u_2(x, y)$ solution of

$$\begin{aligned}
&-\nabla_y \cdot (A(y)\nabla_y u_2) \\
&\quad = \nabla_x \cdot (A(y)\nabla_x u_0) + \nabla_x \cdot (A(y)\nabla_y u_1) + \nabla_y \cdot (A(y)\nabla_x u_1) + f_0 \quad \text{in } Y, \\
&\qquad -A(y)\nabla_y u_2 \cdot n(y) = A(y)\nabla_x u_1 \cdot n(y) + g_0 \quad \text{at } \partial Y_0, \\
&\qquad\qquad u_2 \text{ is } Y\text{-periodic.}
\end{aligned}$$

Observe that (2.57) gives

$$I_1 + I_2 + I_3 = I_4 + I_5,$$

where the integral terms I_k ($k \in \{1, \ldots, 5\}$) are defined by

$$\begin{aligned}
I_1 &:= \int_Y \nabla_x \cdot (A(y)\nabla_x u_0)dy, \\
I_2 &:= \int_Y \nabla_x \cdot (A(y)\nabla_y u_1)dy + \int_Y \nabla_y \cdot (A(y)\nabla_x u_1)dy, \\
I_3 &:= \int_Y f_0 dy, \\
I_4 &:= \int_{\partial Y_0} A(y)\nabla_x u_1 \cdot n(y)d\sigma_y, \\
I_5 &:= \int_{\partial Y_0} g_0 d\sigma_y.
\end{aligned}$$

We have

$$\begin{aligned}
I_1 &= \int_Y \nabla_x \cdot (A(y)\nabla_x u_0)dy \\
&= \nabla_x \cdot \left[\left(\int_Y A(y)dy\right)\nabla_x u_0\right] = \left(\int_Y A(y)dy\right) : \nabla_x \nabla_x u_0(x).
\end{aligned} \tag{2.58}$$

In (2.58), we have used the inner product between two matrices:

$$A : B := \mathrm{tr}(A^t B) = \sum_{ij} a_{ij} b_{ij}.$$

This product,[11] also called the double dot product, is used to represent multiplication and summation across two indices. It is worth noting that

[11] Given vector fields a, v, we denote $a \cdot \nabla v := (\nabla v)a$. Similarly, if A is a second-order tensor, we denote $a \cdot \nabla A := (\nabla A)a$ and understand the notation component-wise. For any scalar function ψ, $\nabla\nabla\psi$ denotes the Hessian matrix. Note also that, since $\Delta\psi = \nabla \cdot \nabla\psi$, we can use the notation $\Delta\psi = \mathbb{I} : \nabla\nabla\psi$, where $\mathbb{I}$ is the identity matrix.

the double dot product of two second-order tensors is a scalar. Some useful properties of the double dot product are listed, for instance, in Pavliotis and Stuart (2008, Section 2.2).

Furthermore, $I_2 = I_{21}+I_{22}$. By combining the periodicity of the involved functions with Gauß's theorem, we obtain

$$I_{22} = I_4.$$

We now shift our focus on the term I_{22}. Concretely, we obtain that

$$\begin{aligned} I_{21} &= \int_Y \nabla_x \cdot \big(A(y)\nabla_y u_1\big)dy = \int_Y A(y) : \nabla_x \nabla_y \big(W(y)\cdot \nabla_x u_0(x)\big)dy \\ &= \int_Y \big(A(y)\nabla_y W(y)^t\big)dy : \nabla_x\nabla_x u_0(x). \end{aligned}$$

Combining the above results yields the following *upscaled problem*:

Find $u_0 \in H_0^1(\Omega)$ satisfying weakly the following equations:

$$\begin{aligned} -\frac{|Y|}{|Z|} &\fint_Y \big(A(y) + A(y)\nabla_y W(y)^t\big)dy : \nabla_x \nabla_x u_0(x) \\ &= \frac{|Y|}{|Z|} f_0(x) + \frac{|\partial Y_0|}{|Z|}\fint_{\partial Y_0} g_0(x,y)d\sigma_y, \end{aligned} \tag{2.59}$$

for a.e. $x \in \Omega$, with

$$u_0(x) = 0 \quad \text{for all } x \in \partial\Omega. \tag{2.60}$$

In (2.59), we denote by $\fint_{\mathcal{Y}} f(y)dy$ the quantity $\frac{1}{|\mathcal{Y}|}\int_{\mathcal{Y}} f(y)d\mu_y \in \mathbb{R}$ for any measurable set with respect to the (volume or surface) measure μ_y. Equation (2.59) can also be obtained by integrating by parts (2.57) with respect to y and then by multiplying the result with $\frac{1}{|Z|}$.

It is important to note that the factor $\frac{|Y|}{|Z|}$ is the so-called *volumetric porosity* and is usually denoted by ϕ, while $\frac{|\partial Y_0|}{|Z|}$ denotes the *surface porosity*. We refer the reader, for instance, to the monograph by Bear (1988) for related porous media concepts arising when modeling flow, diffusion, and chemical reactions in domains with heterogeneities. To emphasize the connection with the theory of porous media, we can rewrite the upscaled problem in a slightly modified form:

$$\mathrm{div}_x(-\mathbb{D}\nabla u_0(x)) = \phi F(x) \quad \text{for } x \in \Omega, \tag{2.61}$$

with

$$u_0(x) = 0 \quad \text{for all } x \in \partial\Omega, \tag{2.62}$$

where the effective diffusion tensor reads

$$\mathbb{D} := \phi \fint_Y (A(y) + A(y)\nabla_y W(y)^t)dy, \tag{2.63}$$

while the effective production term by reaction is

$$\begin{aligned} \phi F(x) &:= \frac{|Y|}{|Z|} f_0(x) + \frac{|\partial Y_0|}{|Z|} \fint_{\partial Y_0} g_0(x,y)d\sigma_y \\ &= \frac{|Y|}{|Z|}\left(f_0(x) + \frac{|\partial Y_0|}{|Y|} \fint_{\partial Y_0} g_0(x,y)d\sigma_y\right). \end{aligned} \tag{2.64}$$

When handling microscopic problems arising from practical scenarios, engineers are often interested in computing or at least estimating the so-called tortuosity tensor, say $\mathbb{T}$. Denoting by $\tilde{A}$ the known value of the diffusion coefficient of the gaseous molecule in air, we recall from Bear (1988) that

$$\mathbb{D} := \phi \tilde{A} \mathbb{T}. \tag{2.65}$$

In this case, the tortuosity tensor has the exact representation

$$\mathbb{T} := \frac{1}{\tilde{A}} \fint_Y (A(y) + A(y)\nabla_y W(y)^t)dy. \tag{2.66}$$

Note that the entries of the tortuosity tensor (2.66) and hence also those of the effective diffusion tensor (2.63) are computable quantities.

2.4 The Cell Problem

It is not *a priori* clear whether the obtained macroscopic problem has a good chance of being well-posed. It turns out that the effective diffusion tensor $\mathbb{D}$ is coercive. This fact ensures the desired (weak) solvability and other properties of the solution. To see this, it is convenient to introduce the Hilbert space

$$\mathcal{W} := \left\{ \psi \in H^1_{\#}(Y) : \int_Y \psi(y)dy = 0 \right\}$$

endowed with the norm

$$\|\psi\|_{\mathcal{W}} := \left(\int_Y |\nabla_y \psi(y)|^2 dy \right)^{1/2}.$$

We recall the structure of the cell problem obtained previously: For all $j \in \{1, \ldots, d\}$, we search for the cell functions $W_j \in H^1_{\#}(Y)$ such that the boundary-value problem (i.e. the cell problem)

$$\nabla_y \cdot \big(- A(y) \nabla_y W_j(y) \big) = \nabla_y \cdot \big(A(y) e_j \big) \quad \text{for } y \in Y, \tag{2.67}$$

$$-A(y) \nabla_y W_j(y) \cdot n(y) = 0 \quad \text{for } y \in \partial Y_0, \tag{2.68}$$

$$W_j \text{ is } Y\text{-periodic} \quad \int_Y W_j dy = 0 \tag{2.69}$$

is satisfied weakly. With this in mind, the weak formulation corresponding to (2.67)–(2.69) reads as follows:

Find $W_j \in \mathcal{W}$ such that

$$\int_Y A(y) \nabla_y W_j(y) \nabla_y \psi(y) dy = - \int_Y A(y) e_j \nabla_y \psi(y) dy \tag{2.70}$$

holds for all $\psi \in \mathcal{W}$.

A direct application of the Lax–Milgram lemma guarantees the existence and uniqueness of a weak solution in the sense of (2.70) if the matrix $A(\cdot)$ is coercive.

We now show that the effective diffusion tensor $\mathbb{D}$ is coercive and, hence, also positive definite. The argument is standard and applies to a large variety of cell problems different from the one mentioned here. Proceeding as outlined by Alouges (2016), we take $\psi = W_k \in \mathcal{W}$ as a test function in (2.70) for some $k \in \{1, \ldots, d\}$. We obtain

$$\int_Y A(y)\big(e_j + \nabla_y W_j(y)\big)\nabla_y W_k(y)dy = 0. \tag{2.71}$$

Hence, the effective diffusion tensor satisfies the identity

$$\begin{aligned}
D_{jk} &= \phi \fint_Y \left(A_{jk}(y) + \sum_{\ell=1}^{d} A_{j\ell}(y) \frac{\partial W_k}{\partial y_\ell}(y) \right) dy, \\
&= \phi \fint_Y \big(A(y)e_k + A(y)\nabla_y W_k(y)\big) e_j dy \\
&= \phi \fint_Y A(y)\big(e_k + \nabla_y W_k(y)\big) e_j dy \\
&= \phi \fint_Y A(y)\big(e_k + \nabla_y W_k(y)\big)\big(e_j + \nabla_y W_j(y)\big) dy,
\end{aligned}$$

where we used (2.71) to obtain the last equality.

By a direct calculation, we see that for all $\xi \in \mathbb{R}^d$, we are led to

$$\begin{aligned}
(\mathbb{D}\xi, \xi)_{\mathbb{R}^d} &= \sum_{j=1}^{d}\sum_{k=1}^{d} D_{jk}\xi_j\xi_k \\
&= \phi \fint_Y A(y) \left(\xi + \sum_{k=1}^{d} \xi_k \nabla_y W_k(y) \right) \left(\xi + \sum_{j=1}^{d} \xi_j \nabla_y W_j(y) \right) dy \\
&\geq c \left\| \xi + \sum_{k=1}^{d} \xi_k \nabla_y W_k(y) \right\|_{L^2(Y)}^2 ,
\end{aligned} \tag{2.72}$$

where $c > 0$ is a constant whose value depends explicitly on ϕ and on the coercivity constant of the matrix $A(\cdot)$. The notation $(\cdot, \cdot)_{\mathbb{R}^d}$ arising in (2.72) refers to the usual scalar product in $\mathbb{R}^d$.

It is straightforward to see that if the matrix $A(\cdot)$ is symmetric, then so is the effective tensor $\mathbb{D}$.

2.5 Imperfect Approximations: Higher-Order Homogenization Asymptotics

The homogenization procedure detailed in Section 2.3.4 requires the use of hypercube-type domains Ω, and it seems to be closed without any more assumptions than those mentioned in the asymptotic *ansatz* (2.37). At this stage, we only wish to mention that the domain Ω does not have to be *per se* a hypercube, but it should be approximated sufficiently well by a suitable union of hypercubes. Typical situations are described by Ciorănescu and Saint Jean Paulin (1998). Moreover, in some cases, it is possible to perform the so-called "partial homogenization", i.e. asymptotic expansions are now prepared to encode different type of oscillations (or even no oscillations at all) in different parts of the domain Ω. For more on this matter, we refer the reader to Panasenko (2002). The two-scale asymptotics expansion is indeed close if only the first term in the expansion is our focus. However, there are many examples of applications of the homogenization method to real-world problems where the first order in the expansion, i.e. $u_0(x)$, does not capture interesting effects; see, for instance, the derivation of a continuum model for hierarchical fibril assembly reported by van Lith *et al.* (2014). This means that interesting effects contained in "good" fluctuations are likely to be hidden within the higher orders of the expansion. The calculations done in Section 2.3.4 indicate that there are hierarchical structures that can be employed to unveil information about subsequent orders u_1, u_2, and so on, possibly involving more classes of cell problems. The reader will find in the work of Ciorănescu and Saint Jean Paulin (1998) how higher-order expansions deliver information on the speed of convergence of the homogenization process (i.e. on the structure of the correctors).

2.6 Memory Effects

Sometimes averaging procedures introduce memory effects visible at the level of upscaled equations as nonlocal terms (Du *et al.*, 2020). As time elapses, memory terms may or may not fade depending very much on the model parameters. If one thinks of memory effects arising when applying homogenization procedures, a standard example is the following model problem initially proposed by L. Tartar. Let us assume for the moment that the set $\emptyset \neq \Omega \subset \mathbb{R}^d$ is open and that the coefficient $a_\varepsilon \in L^\infty(\Omega)$, with $0 < \alpha \leq a_\varepsilon(x) \leq \beta < \infty$ for a.e. $x \in \Omega$, converges only weakly to some function a_0. Additionally, let $f_\varepsilon \in C([0,T] \times \bar{\Omega})$ by converging to some function f_0. To fix ideas, take for instance $f_\varepsilon(t,x) := \varepsilon + f_0(t,x)$ for $(t,x) \in [0,T] \times \bar{\Omega}$.

Let u_ε satisfy the following linear equation with oscillating coefficients:

$$\partial_t u_\varepsilon(t,x) + a_\varepsilon(x) u_\varepsilon(t,x) = f_\varepsilon(t,x) \quad \text{in } (0,T) \times \Omega, \tag{2.73}$$

with $u_\varepsilon(x,0) = 0$ for $x \in \Omega$. One can show (see Theorem 1.2 in Tartar (1990)) that, up to a subsequence, u_ε convergences weakly to some function u_0 solution of the following integro-differential equation:

$$\partial_t u_0(t,x) + a_0(x)u_0(t,x) - \int_0^t K(x,t-s)u_0(s,x)ds = f_0(t,x) \quad \text{in } (0,T)\times\Omega, \tag{2.74}$$

with $u_0(x,0) = 0$ for $x \in \Omega$. The resulting kernel K is measurable in $x \in \Omega$ and analytic in $t \in (0,T)$. Furthermore, K admits the representation

$$K(x,t) = \int e^{-t\lambda} d\mu_x(\lambda) \quad \text{a.e. } x \in \Omega.$$

The parameter-dependent measure $\mu_x(\lambda)$ – the so-called Young measure – depends on the structure of the oscillating function a_ε. One can see this fact easily by means of the same example (2.75) when taking $f_\varepsilon(t,x) = -a_\varepsilon(x)h(x)$. We take here $h \in L^2(\Omega)$ and a_ε as before. Hence, it holds that $f_\varepsilon \in C([0,T], L^2(\Omega))$. The new problem reads as follows: Find W_ε such that

$$\partial_t W_\varepsilon(t,x) + a_\varepsilon(x)W_\varepsilon(t,x) = -a_\varepsilon(x)h(x) \quad \text{in } (0,T)\times\Omega, \tag{2.75}$$

with $W_\varepsilon(x,0) = 0$ for $x \in \Omega$. Setting now

$$r_\varepsilon(t,x) := W_\varepsilon(t,x) + h(x)$$

leads to the following equation: Find r_ε such that

$$\partial_t r_\varepsilon(t,x) + a_\varepsilon(x)r_\varepsilon(t,x) = 0 \quad \text{in } (0,T)\times\Omega, \tag{2.76}$$

with $r_\varepsilon(x,0) = h(x)$ for $x \in \Omega$. For any fixed $\varepsilon > 0$, this equation admits the solution

$$r_\varepsilon(x,t) = h(x)e^{-ta_\varepsilon(x)}.$$

Arguing as done by Du *et al.* (2020), as (a_ε) convergences only weakly when $\varepsilon \to 0$, the weak limit r_0 of (r_ε) cannot be expressed usually as $e^{-ta_0(x)}$. In fact, using the concept of the Young measure, the weak limit r_0 becomes

$$r_0(t,x) = h(x)\int e^{-t\lambda} d\mu_x(\lambda) \quad \text{a.e. } x \in \Omega \text{ and all } t \in [0,T).$$

For specific simple choices of a_ε, the measure μ_x can be computed explicitly.

2.7 Exercises

Exercise 2.7.1. Apply the Lax–Milgram lemma to prove the existence and uniqueness of $w_\varepsilon \in H_0^1(\Omega)$ satisfying weakly the boundary-value problems (2.5)–(2.6), and similarly for $p_\varepsilon \in H_0^1(\Omega)$ satisfying weakly the boundary-value problems (2.8)–(2.9).

Exercise 2.7.2. Prove the ε-independent energy estimates (2.13) and (2.14).

Exercise 2.7.3. Let $a \in L_+^\infty(\Omega)$ be a 1-periodic function. The following statements are equivalent:

(i) $\frac{1}{\langle \frac{1}{a} \rangle} = \langle a \rangle$;
(ii) a is a constant.

Exercise 2.7.4. Is the periodicity assumption on a_ε essential to perform the homogenization asymptotics? Justify your answer.

Exercise 2.7.5. Concerning the diffusion through the periodically laminated material described earlier, complete the mentioned homework, prove that $\frac{\partial \sigma_1^\epsilon}{\partial x_1} \in L_{x_1}^2(H_{x_2}^{-1})$, and show also the following convergences as $\epsilon \to 0$ (holding in appropriate spaces that you must indicate):

$$\sigma_\epsilon \to \sigma_1$$

$$\frac{1}{a(x_1/\epsilon)}\sigma_\epsilon \rightharpoonup \left\langle \frac{1}{a} \right\rangle \sigma_1$$

$$a(x_1/\epsilon)u_\epsilon \rightharpoonup \langle a \rangle u_0.$$

Exercise 2.7.6. Prove (2.45).

Exercise 2.7.7. Prove (2.49). Additionally, show why Lemma 7.1 does actually apply in this context.

Exercise 2.7.8. Apply the Lax–Milgram lemma to (2.70).

Exercise 2.7.9. Let ε be a small positive parameter. On the set $\Omega := (0,1)^2$, we define the ε-periodic function $a_\varepsilon(x) := a(x_1/\varepsilon)$ depending only on the first coordinate x_1 so that it satisfies

$$0 < m \le a_\epsilon(x) \le M < \infty$$

for some given (independent of ε) constants m and M. Let $q_\varepsilon(x) := q(x/\varepsilon)$ be a given ε-periodic velocity vector (referred to as *drift*) satisfying the divergence-free condition

$$\operatorname{div}(q_\varepsilon) = 0.$$

For a given forcing term f, we consider the microscopic equation

$$\operatorname{div}\big(q_\varepsilon u^\varepsilon - a_\varepsilon \nabla u^\varepsilon\big) = f \quad \text{in } \Omega,$$

with homogeneous Dirichlet boundary conditions. Apply the two-scale asymptotic expansion to determine the structure of the upscaled equation. Determine the explicit structure of the effective diffusion tensor, as well as the one for the effective drift.

Exercise 2.7.10. Consider the setting described in the previous exercise. Additionally, assume $a_\varepsilon \in L^\infty(\Omega)$ and $f \in L^2(\Omega)$. What regularity (and eventual other assumptions) you need for q_ε so that you can prove rigorously the passage to the homogenization limit? Show the complete proof of the derivation of the upscaled equation, as well as the structure of the effective diffusion coefficient and the corresponding effective drift.

Exercise 2.7.11. Solve Exercises 1c, 2, and 4 from Holmes (1995, pp. 231–233).

Exercise 2.7.12. We bring to the reader's attention the text in Hornung (1997, pp. 18–22). It deals with the formal derivation of a linear transmission problem for the system of stationary diffusion equations, where the two diffusion coefficients, say D_1 and D_2, satisfy the very special scaling (in ε) $D_1 = \varepsilon^2 D_2$. The resulting system of equations is usually referred in the literature as "double porosity" or the "distributed microstructure" model. Redo the calculations from Hornung (1997) (loc. cit.) for the case when $\frac{D_1}{D_2} = \mathcal{O}(1)$.

Exercise 2.7.13. Consider as a microscopic target problem the linear time-dependent diffusion-convection equation,

$$\partial_t u_\varepsilon + \operatorname{div}(-a_\varepsilon \nabla u_\varepsilon + b_\varepsilon u_\varepsilon) = f_\varepsilon \quad \text{in } \Omega_\varepsilon \subset \mathbb{R}^d,$$

together with the boundary conditions

$$u_\varepsilon = 0 \quad \text{on } \Gamma^{ext},$$

$$(-a_\varepsilon \nabla u_\varepsilon + b_\varepsilon u_\varepsilon) \cdot n_\varepsilon = g_\varepsilon \quad \text{on } \Gamma_\varepsilon$$

and the prescribed initial condition

$$u_\varepsilon(t=0,x) = u_I^\varepsilon(x) \quad \text{for all } x \in \bar{\Omega}_\varepsilon.$$

As customary for this type of problems, all data are Y-periodic. The notation of the geometry of Ω_ε is as discussed in Section 2.3.2. Besides assuming a uniform ellipticity for a_ε and everything else that is needed for the microscopic problem to be well-posed (in a weak sense), we impose

$$f_\varepsilon(x) = f_0(x) + \varepsilon^2 f_1\left(x, \frac{x}{\varepsilon}\right),$$

$$g_\varepsilon(x) = \varepsilon g_0\left(x, \frac{x}{\varepsilon}\right) + \varepsilon^2,$$

as well as the incompressibility condition

$$\operatorname{div}(b_\varepsilon) = 0.$$

The main task is twofold: Perform the formal two-scale homogenization procedure for this problem, and then derive the macroscopic model equations as well as the calculation rules for the effective diffusion coefficient and the effective drift. Two additional questions can be addressed in this context:

- What do you need to assume on the initial data $u_I^\varepsilon(x)$ to perform the formal homogenization?
- Is the assumption $\operatorname{div}(b_\varepsilon) = 0$ essential to perform the homogenization procedure? How would you perform it if the flow, with velocity b_ε, is in fact not incompressible?

Exercise 2.7.14. Take the production term f_ε from the previous example and assume additionally that $g \in C^\infty(\mathbb{R} \times \mathbb{R})$. Let $\alpha \in \mathbb{R}$ be a parameter fixed arbitrarily. Consider the following microscopic problem: Find $(u_\varepsilon, v_\varepsilon)$ such that

$$\partial_t u_\varepsilon + \operatorname{div}(-D_\varepsilon \nabla u_\varepsilon) = f_\varepsilon \quad \text{in } \Omega_\varepsilon,$$

together with the boundary conditions

$$u_\varepsilon = 0 \quad \text{on } \Gamma^{ext},$$

$$-D_\varepsilon \nabla u_\varepsilon \cdot n_\varepsilon = \varepsilon^\alpha \partial_t v_\varepsilon \quad \text{on } \Gamma_\varepsilon,$$

$$\partial_t v_\varepsilon = g(u_\varepsilon, v_\varepsilon) \quad \text{on } \Gamma_\varepsilon$$

and the initial conditions

$$u_\varepsilon(t=0,x) = u^0(x) \quad \text{for } x \in \bar{\Omega}_\varepsilon,$$
$$v_\varepsilon(t=0,x) = v^0(x) \quad \text{for } x \in \Gamma_\varepsilon.$$

Identify the upscaled equations depending on the choice of α. Are there ranges of the parameter α for which you cannot perform the homogenization procedure?

Exercise 2.7.15. Take the same setting as in the previous exercise. Moreover, let b_ε be an incompressible vector field and take $\beta \in \mathbb{R}$ as an arbitrary parameter. Consider the following Fokker–Planck-type microscopic problem: Find u_ε such that

$$\partial_t u_\varepsilon - \Delta(D_\varepsilon u_\varepsilon) + \varepsilon^\alpha \nabla u_\varepsilon \cdot b_\varepsilon = f_\varepsilon \quad \text{in } \Omega_\varepsilon,$$

together with the boundary conditions

$$u_\varepsilon = 0 \quad \text{on } \Gamma^{ext},$$
$$(-\nabla(D_\varepsilon u_\varepsilon) + \varepsilon^\alpha u_\varepsilon b_\varepsilon) \cdot n_\varepsilon = \varepsilon^\beta \partial_t u_\varepsilon \quad \text{on } \Gamma_\varepsilon$$

and the initial condition

$$u_\varepsilon(t=0,x) = u^0(x) \quad \text{for } x \in \bar{\Omega}_\varepsilon.$$

Identify the upscaled equations depending on the choices of α and β. Are there ranges of the parameters α and β for which you cannot perform the homogenization procedure?

Exercise 2.7.16. Consider the following coupled system of partial differential equations: Find the pair $(m_\varepsilon, \phi_\varepsilon)$ such that the system

$$\partial_t m_\varepsilon = \operatorname{div}(-A_\varepsilon \nabla m_\varepsilon - 2\varepsilon^\gamma \beta_\varepsilon(\phi_\varepsilon - m_\varepsilon^2)\nabla J \star m_\varepsilon),$$
$$\partial_t \phi_\varepsilon = \operatorname{div}(-B_\varepsilon \nabla \phi_\varepsilon - 2\varepsilon^\gamma \beta_\varepsilon(1 - \phi_\varepsilon)\nabla J \star m_\varepsilon)$$

holds in $\Omega \times (0,T)$, together with periodic boundary conditions and the initial conditions

$$m_\varepsilon(t=0) = m_\varepsilon^0, \quad \phi_\varepsilon(t=0) = \phi_\varepsilon^0.$$

Here, J is a suitable smooth potential. The notation $\nabla J \star m_\varepsilon$ indicates that ∇J is convoluted with the function m_ε, while $A_\varepsilon(x)$, $B_\varepsilon(x)$, and $\beta_\varepsilon(x)$ are, in fact, $A\left(\frac{x}{\varepsilon}\right)$, $B\left(\frac{x}{\varepsilon}\right)$, and $\beta\left(\frac{x}{\varepsilon}\right)$, respectively. These coefficients are all Y-periodic. The data and parameters are chosen so that the model is well-posed. Identify the upscaled equations depending on the choice of γ. Are there ranges concerning the parameter γ for which you cannot perform the homogenization procedure? Justify your answer.

2.8 Solutions

Solution 2.7.3. We note that

$$\langle a \rangle \left\langle \frac{1}{a} \right\rangle \geq 1.$$

Solution 2.7.4. The answer is negative. The periodicity assumption on the coefficient a_ε was used to identify $\langle a \rangle$ as the corresponding weak-$\star$ limit in $L^\infty(\Omega)$. If we relax this constraint on a_ε while keeping all the other assumptions unchanged, we can still pass to the limit and identify the structure of the limit problem, but we generally loose the uniqueness of solutions (think of situations involving high-contrast composites, like the cheeso-steel case mentioned in Chapter 1); see also Remark 1.10 in Bensoussan *et al.* (1978). However, if the sequence (a_ε) admits an unique accumulation point, then the uniqueness of solutions is restored.

Solution 2.7.5. Let $X_1 := (0,1)$ and $X_2^\delta := (0,\delta)$, and hence, $\Omega^\delta = X_1 \times X_2^\delta$. We denote $L^2_{x_1} := L^2(X_1)$, $L^2_{x_2} := L^2(X_2^\delta)$, $H^1_{x_1} := H^1(X_1)$, $H^1_{x_2} := H^1(X_2^\delta)$, and $L^2_{x_1}(H^{-1}_{x_2}) := L^2(X_1; H^{-1}(X_2^\delta))$. We perceive $L^2(\Omega^\delta)$ as $L^2_{x_1} \otimes L^2_{x_2}$ and $H^1(\Omega^\delta)$ as $H^1_{x_1} \otimes H^1_{x_2}$; we rely on the structure of the underlying function spaces and on a suitable use of Fubini's theorem.

We plan to show that

$$\frac{\partial \sigma_1^\varepsilon}{\partial x_1} = f - \frac{\partial \sigma_2^\varepsilon}{\partial x_2} \in L^2(X_1; H^{-1}(X_2^\delta)) \tag{2.77}$$

and that there exists a sequence (ε_n) such that $\varepsilon_n \to 0$ and

$$\sigma_1^{\varepsilon_n} \to \sigma_1 \quad \text{strongly in } L^2(X_1; H^{-1}(X_2^\delta)), \tag{2.78}$$

$$\frac{1}{a_{\varepsilon_n}} \sigma_1^{\varepsilon_n} \rightharpoonup \left\langle \frac{1}{a} \right\rangle \sigma_1 \quad \text{weakly in } L^2(X_1; H^{-1}(X_2^\delta)), \tag{2.79}$$

$$a_{\varepsilon_n} u_{\varepsilon_n} \rightharpoonup \langle a \rangle u_0 \quad \text{weakly in } L^2(\Omega^\delta), \tag{2.80}$$

as $n \to \infty$, where $a_\varepsilon(x_1, x_2) := a(x_1/\varepsilon)$.

We proceed in four steps.

Step 1: We show (2.77) and that the sequence $(\frac{\partial\sigma_1^\varepsilon}{\partial x_1})$ is bounded in $L^2(X_1; H^{-1}(X_2^\delta))$. Since $f, \sigma_2^\varepsilon \in L^2(\Omega^\delta)$, we have $f_{x_1} := f(x_1, \cdot), \sigma_{2,x_1}^\varepsilon := \sigma_2^\varepsilon(x_1, \cdot) \in L^2(X_2^\delta)$ for a.e. $x_1 \in X_1$, and for all $\psi \in H_0^1(X_2^\delta)$ and a.e. $x_1 \in X_1$,

$$\left|\left\langle f_{x_1} - \frac{\partial\sigma_{2,x_1}^\varepsilon}{\partial x_2}, \psi\right\rangle_{H_0^1(X_2^\delta)}\right| = \left|\int_{X_2^\delta}\left(f_{x_1}\psi + \sigma_{2,x_1}^\varepsilon \frac{d\psi}{dx_2}\right)dx_2\right|$$
$$\leq (\|f_{x_1}\|_{L^2(X_2^\delta)} + \|\sigma_{2,x_1}^\varepsilon\|_{L^2(X_2^\delta)})\|\psi\|_{H^1(X_2^\delta)}.$$

Hence, the following inequality holds for a.e. $x_1 \in X_1$:

$$\left\|f_{x_1} - \frac{\partial\sigma_{2,x_1}^\varepsilon}{\partial x_2}\right\|_{H^{-1}(X_2^\delta)} := \sup_{0 \neq \psi \in H_0^1(X_2^\delta)} \frac{\left|\left\langle f_{x_1} - \frac{\partial\sigma_{2,x_1}^\varepsilon}{\partial x_2}, \psi\right\rangle\right|}{\|\psi\|_{H^1(X_2^\delta)}}$$
$$\leq \|f_{x_1}\|_{L^2(X_2^\delta)} + \|\sigma_{2,x_1}^\varepsilon\|_{L^2(X_2^\delta)}.$$

Since

$$\|f_{x_1}\|_{L^2(X_2^\delta)}, \|\sigma_{2,x_1}^\varepsilon\|_{L^2(X_2^\delta)} \in L^2(X_1),$$

we obtain

$$\left\|f_{x_1} - \frac{\partial\sigma_{2,x_1}^\varepsilon}{\partial x_2}\right\|_{H^{-1}(X_2^\delta)} \in L^2(X_1)$$

and

$$\left\|f - \frac{\partial\sigma_2^\varepsilon}{\partial x_2}\right\|_{L^2(X_1;H^{-1}(X_2^\delta))} = \sqrt{\int_{X_1}\left\|f_{x_1} - \frac{\partial\sigma_{2,x_1}^\varepsilon}{\partial x_2}\right\|_{H^{-1}(X_2^\delta)}^2 dx_1}$$
$$\leq \sqrt{\int_{X_1}(\|f_{x_1}\|_{L^2(X_2^\delta)} + \|\sigma_{2,x_1}^\varepsilon\|_{L^2(X_2^\delta)})^2 dx_1}$$
$$\leq \sqrt{2}(\|f\|_{L^2(X_1;L^2(X_2^\delta))} + \|\sigma_2^\varepsilon\|_{L^2(X_1;L^2(X_2^\delta))})$$
$$= \sqrt{2}(\|f\|_{L^2(\Omega^\delta)} + \|\sigma_2^\varepsilon\|_{L^2(\Omega^\delta)}).$$

Furthermore, since we have

$$\|\sigma_2^\varepsilon\|_{L^2(\Omega^\delta)} \leq \|a_\varepsilon\|_{L^\infty(X_1)}\left\|\frac{\partial u_\varepsilon}{\partial x_2}\right\|_{L^2(\Omega^\delta)} \leq \frac{c_2 c_P}{c_1}\|f\|_{L^2(\Omega^\delta)}$$

for all $\varepsilon > 0$, $\frac{\partial\sigma_1^\varepsilon}{\partial x_1} = f - \frac{\partial\sigma_2^\varepsilon}{\partial x_2}$ is bounded in $L^2_{x_1}(H^{-1}_{x_2}) = L^2(X_1; H^{-1}(X_2^\delta))$.

Step 2: Recall that there exist a sequence (ε_n), as well as three functions $u_0 \in L^2(\Omega^\delta)$, $\langle a \rangle \in L^\infty(\Omega^\delta)$, and $\langle \frac{1}{a} \rangle \in L^\infty(X_1)$, such that $\varepsilon_n \to 0$ and

$$u_{\varepsilon_n} \to u_0 \quad \text{strongly in } L^2(\Omega^\delta),$$

$$\nabla u_{\varepsilon_n} \rightharpoonup \nabla u_0 \quad \text{weakly in } L^2(\Omega^\delta),$$

$$a_{\varepsilon_n} (= a_{\varepsilon_n} 1_{X_2^\delta}) \overset{*}{\rightharpoonup} \langle a \rangle \quad \text{weakly-}* \text{ in } L^\infty(\Omega^\delta),$$

$$\frac{1}{a_{\varepsilon_n}} \overset{*}{\rightharpoonup} \left\langle \frac{1}{a} \right\rangle \quad \text{weakly-}* \text{ in } L^\infty(X_1),$$

as $n \to \infty$. This is basically a consequence of a combination of facts, including $\|\nabla u_\varepsilon\|_{L^2(\Omega^\delta)} \leq \frac{c_P}{c_1} \|f\|_{L^2(\Omega^\delta)}$ for all $\varepsilon > 0$, $0 < c_1 \leq a_\varepsilon \leq c_2 < \infty$ on $\mathbb{R}$, and Theorem 7.5 (the Rellich–Kondrachov theorem). Furthermore, from the fourth convergence, it yields that

$$\frac{1}{a_{\varepsilon_n}} 1_{X_2^\delta} \overset{*}{\rightharpoonup} \left\langle \frac{1}{a} \right\rangle 1_{X_2^\delta} \quad \text{weakly-}* \text{ in } L^\infty(\Omega^\delta) \tag{2.81}$$

as $n \to \infty$.[12] Indeed, for all $\varphi \in L^1(\Omega^\delta) \cong L^1(X_1; L^1(X_2^\delta))$, we have $\varphi_{x_1} := \varphi(x_1, \cdot) \in L^1(X_2^\delta)$ for a.e. $x_1 \in X_1$ and

$$\int_{\Omega^\delta} \left(\frac{1}{a_{\varepsilon_n}} 1_{X_2^\delta} - \left\langle \frac{1}{a} \right\rangle 1_{X_2^\delta} \right) \varphi dx = \int_{X_1} \left(\frac{1}{a_{\varepsilon_n}} - \left\langle \frac{1}{a} \right\rangle \right) \left\{ \int_{X_2^\delta} \varphi_{x_1} dx_2 \right\} dx_1 \to 0$$

as $n \to \infty$, where we have used the fact that $\int_{X_2^\delta} \varphi_{x_1} dx_2 \in L^1(X_1)$ for a.e. $x_1 \in X_1$.

Next, we use two known compactness results to prove (2.78), that is, we make use of Lemmas 7.6 and 7.7.

It is essential to note that $L^2(\Omega^\delta)$ is compactly embedding in $H^{-1}(\Omega^\delta)$. By the Rellich–Kondrachov theorem (Theorem 7.5), the inclusion map

$$I : H_0^1(\Omega^\delta) \ni v \mapsto v \in L^2(\Omega^\delta)$$

is compact. Here, we have $H^{-1}(\Omega^\delta) = (H_0^1(\Omega^\delta))^*$ and $L^2(\Omega^\delta) = (L^2(\Omega^\delta))^*$, and the dual operator $I^* : L^2(\Omega^\delta) \to H^{-1}(\Omega^\delta)$ satisfies that, for all $v \in H_0^1(\Omega^\delta)$ and $w \in L^2(\Omega^\delta)$, the expression

$$\langle I^* w, v \rangle_{H_0^1(\Omega^\delta)} = \langle Iv, w \rangle = \int_{\Omega^\delta} vw dx.$$

[12] We simply denote $\frac{1}{a_{\varepsilon_n}} 1_{X_2^\delta}$ and $\langle \frac{1}{a} \rangle 1_{X_2^\delta}$ by $\frac{1}{a_{\varepsilon_n}}$ and $\langle \frac{1}{a} \rangle$, respectively.

This implies I^* is the usual inclusion map of $L^2(\Omega^\delta)$ into $H^{-1}(\Omega^\delta)$. By Lemma 7.6, the inclusion map

$$I^* : L^2(\Omega^\delta) \to H^{-1}(\Omega^\delta)$$

is compact.

Since $(\sigma_1^{\varepsilon_n})_{n\in\mathbb{N}}$ is bounded in $L^2(\Omega^\delta) \cong L^2_{x_1}(L^2_{x_2}) = L^2(X_1; L^2(X_2^\delta))$ and $(\frac{\partial\sigma_1^{\varepsilon_n}}{\partial x_1})_{n\in\mathbb{N}}$ is bounded in $L^2_{x_1}(H^{-1}_{x_2}) = L^2(X_1; H^{-1}(X_2^\delta))$, by setting $B_0 := L^2(X_2^\delta)$, $B_1 := B_2 := H^{-1}(X_d)$, and $p = r := 2$ in Lemma 7.7, there exist a subsequence $(\sigma_1^{\varepsilon_{n(k)}})_{k\in\mathbb{N}}$ and $\sigma_1 \in L^2(\Omega^\delta) = L^2_{x_1}(L^2_{x_2})$ such that

$$\begin{aligned}
\sigma_1^{\varepsilon_{n(k)}} &\to \sigma_1 \quad \text{strongly in } L^2_{x_1}(H^{-1}_{x_2}),\\
\sigma_1^{\varepsilon_{n(k)}} &\rightharpoonup \sigma_1 \quad \text{weakly in } L^2_{x_1}(L^2_{x_2})
\end{aligned}$$

as $k \to \infty$ (and hence, $\varepsilon_{n(k)} \to 0$).

Now, we show (2.79) and (2.80). For simplicity, we denote $\varepsilon_{n(k)}$ by ε_k.

Step 3: For all $\varphi \in L^2(X_1; H_0^1(X_2^\delta))$, it holds that

$$\begin{aligned}
&\left|\int_{X_1} \left\langle \left(\frac{1}{a_{\varepsilon_k}}\sigma_{1,x_1}^{\varepsilon_k} - \left\langle \frac{1}{a}\right\rangle \sigma_{1,x_1}\right), \varphi_{x_1}\right\rangle dx_1\right| \\
&\qquad \le \left|\int_{X_1} \frac{1}{a_{\varepsilon_k}} \langle \sigma_{1,x_1}^{\varepsilon_k} - \sigma_{1,x_1}, \varphi_{x_1}\rangle\, dx_1\right| \\
&\qquad\quad + \left|\int_{X_1} \left(\frac{1}{a_{\varepsilon_k}} - \left\langle\frac{1}{a}\right\rangle\right) \langle \sigma_{1,x_1}, \varphi_{x_1}\rangle\, dx_1\right|.
\end{aligned}$$

Since a_ε is bounded from below by $c_1 > 0$ and $\sigma_1^{\varepsilon_k} \to \sigma_1$ strongly in $L^2_{x_1}(H^{-1}_{x_2})$ as $k \to \infty$, we have that

$$\begin{aligned}
&\left|\int_{X_1} \frac{1}{a_{\varepsilon_k}} \langle \sigma_{1,x_1}^{\varepsilon_k} - \sigma_{1,x_1}, \varphi_{x_1}\rangle\, dx_1\right| \\
&\qquad \le \frac{1}{c_1}\int_{X_1} \left|\langle \sigma_{1,x_1}^{\varepsilon_k} - \sigma_{1,x_1}, \varphi_{x_1}\rangle\right| dx_1 \\
&\qquad \le \frac{1}{c_1}\int_{X_1} \left\|\sigma_{1,x_1}^{\varepsilon_k} - \sigma_{1,x_1}\right\|_{H^{-1}(X_2^\delta)} \left\|\varphi_{x_1}\right\|_{H^1(X_2^\delta)} dx_1 \\
&\qquad \le \frac{1}{c_1}\left\|\sigma_{1,x_1}^{\varepsilon_k} - \sigma_{1,x_1}\right\|_{L^2(X_1;H^{-1}(X_2^\delta))} \left\|\varphi\right\|_{L^2(X_1;H^1(X_2^\delta))}
\end{aligned}$$

tends to zero as $k \to \infty$. On the other hand, since $\langle \sigma_{1,x_1}, \varphi_{x_1} \rangle \in L^1(X_1)$ and $\frac{1}{a_{\varepsilon_k}} \overset{*}{\rightharpoonup} \langle \frac{1}{a} \rangle$ weakly-$*$ in $L^\infty(X_1)$ as $k \to \infty$,

$$\left| \int_{X_1} \left(\frac{1}{a_{\varepsilon_k}} - \left\langle \frac{1}{a} \right\rangle \right) \langle \sigma_{1,x_1}, \varphi_{x_1} \rangle \, dx_1 \right| \to 0$$

as $k \to \infty$.

Step 4: Since $u_{\varepsilon_k} \to u_0$ strongly in $L^2(\Omega^\delta)$ and $a_{\varepsilon_k} \overset{*}{\rightharpoonup} \langle a \rangle$ weakly-$*$ in $L^\infty(\Omega^\delta)$ as $k \to \infty$, it results that for all $\varphi \in L^2(\Omega^\delta)$, the upper bound

$$\begin{aligned} \left| \int_{\Omega^\delta} (a_{\varepsilon_k} u_{\varepsilon_k} - \langle a \rangle u_0) \varphi dx \right| &\leq \left| \int_{\Omega^\delta} (a_{\varepsilon_k} u_{\varepsilon_k} - a_{\varepsilon_k} u_0) \varphi dx \right| \\ &\quad + \left| \int_{\Omega^\delta} (a_{\varepsilon_k} u_0 - \langle a \rangle u_0) \varphi dx \right| \\ &\leq \|a_{\varepsilon_k}\|_{L^\infty(\Omega^\delta)} \|u_{\varepsilon_k} - u_0\|_{L^2(\Omega^\delta)} \|\varphi\|_{L^2(\Omega^\delta)} \\ &\quad + \left| \int_{\Omega^\delta} (a_{\varepsilon_k} - \langle a \rangle) u_0 \varphi dx \right| \end{aligned}$$

vanishes as $k \to \infty$. Here, we have also used the fact that $u_0 \varphi \in L^1(\Omega^\delta)$. Finally, we obtain (2.80).

Solution 2.7.10. We proceed similarly as for the layered medium example and show that the desired homogenized equation is of the following form:

$$\begin{aligned} &- \left(\frac{1}{|Y|} \int_Y a(y)(\mathbb{I} + \nabla_y W(y)^t) dy \right) : \nabla_x \nabla_x u_0(x) \\ &+ \left(\frac{1}{|Y|} \int_Y (\mathbb{I} + \nabla_y W(y)) q(y) dy \right) \cdot \nabla_x u_0(x) = f(x), \end{aligned}$$

where $\mathbb{I}$ is the identity matrix.

Inserting the homogenization *ansatz*

$$u_\varepsilon(x) := u_0(x,y) + \varepsilon u_1(x,y) + \varepsilon^2 u_2(x,y) + \mathcal{O}(\varepsilon^2)|_{\text{where } y:=x/\varepsilon}$$

into the microscopic equation yields

$$\begin{aligned} &\left(\nabla_x + \frac{1}{\varepsilon} \nabla_y \right) \cdot \left[q(y) \left(u_0 + \varepsilon u_1 + \varepsilon^2 u_2 + \text{h.o.t} \right) \right] \\ &\quad + \left(\nabla_x + \frac{1}{\varepsilon} \nabla_y \right) \cdot \left[-a(y) \left(\nabla_x + \frac{1}{\varepsilon} \nabla_y \right) \left(u_0 + \varepsilon u_1 + \varepsilon^2 u_2 + \text{h.o.t} \right) \right] = f(x). \end{aligned}$$

Hence, after some calculations, we obtain

$$\begin{aligned}
&-\frac{1}{\varepsilon^2}\nabla_y \cdot (a(y)\nabla_y u_0)\\
&\quad+\frac{1}{\varepsilon}\left[\nabla_y \cdot (q(y)u_0) - \nabla_y \cdot (a(y)\nabla_x u_0 + a(y)\nabla_y u_1) - \nabla_x \cdot (a(y)\nabla_y u_0)\right]\\
&\quad+\varepsilon^0\left[\nabla_y \cdot (q(y)u_1) + \nabla_x \cdot (q(y)u_0) - \nabla_y \cdot (a(y)\nabla_x u_1 + a(y)\nabla_y u_2)\right.\\
&\quad\left.-\nabla_x \cdot (a(y)\nabla_x u_0 + a(y)\nabla_y u_1)\right] + \text{h.o.t}\\
&= f(x).
\end{aligned}$$

Consequently, we get in $\Omega \times Y$ the following boundary-value problems posed in Y, with $x \in \Omega$ as the parameter:

$$\begin{cases} \nabla_y \cdot (a(y)\nabla_y u_0) = 0, \\ u_0(x,\cdot) \text{ is } Y\text{-periodic for each given } x \in \Omega, \end{cases} \tag{2.82}$$

$$\begin{cases} \nabla_y \cdot (q(y)u_0) - \nabla_y \cdot (a(y)\nabla_x u_0 + a(y)\nabla_y u_1) - \nabla_x \cdot (a(y)\nabla_y u_0) = 0, \\ u_1(x,\cdot) \text{ is } Y\text{-periodic for each given } x \in \Omega, \end{cases} \tag{2.83}$$

as well as

$$\begin{cases} \nabla_y \cdot (q(y)u_1) + \nabla_x \cdot (q(y)u_0) \\ \quad - \nabla_y \cdot (a(y)\nabla_x u_1 + a(y)\nabla_y u_2) - \nabla_x \cdot (a(y)\nabla_x u_0 + a(y)\nabla_y u_1) = f(x), \\ u_2(x,\cdot) \text{ is } Y\text{-periodic for each given } x \in \Omega. \end{cases} \tag{2.84}$$

From (2.82), we note that u_0 is independent of the choice of y, that is,

$$u_0(x,y) = u_0(x) \quad \text{for } (x,y) \in \Omega \times Y. \tag{2.85}$$

Hence, by the divergence-free condition and (2.85), we obtain for all $y \in Y$ that

$$\begin{aligned} \nabla_y \cdot (q(y)u_0) = (\nabla_y \cdot q(y))u_0 + q(y) \cdot \nabla_y u_0 = 0, \\ \nabla_x \cdot (a(y)\nabla_y u_0) = 0, \end{aligned}$$

and (2.83) can be written as

$$\begin{cases} -\nabla_y \cdot (a(y)\nabla_x u_0 + a(y)\nabla_y u_1) = 0, \\ u_1(x,\cdot) \text{ is } Y\text{-periodic for each given } x \in \Omega. \end{cases} \tag{2.86}$$

Here, we use the cell functions $W(y) = (W_1(y), W_2(y))^t$ which satisfy the cell problem

$$\begin{cases} -\nabla_y \cdot (a(y)e_i + a(y)\nabla_y W_i(y))) = 0 \quad \text{for all } y \in Y, \\ W_i \text{ is } Y\text{-periodic}, \ \displaystyle\int_Y W_i \mathrm{d}y = 0, \end{cases}$$

for all $i \in \{1, 2\}$. We set

$$\tilde{u}_1(x,y) := u_1(x,y) - W(y) \cdot \nabla_x u_0(x).$$

Since we have

$$\begin{aligned} &\nabla_y \cdot (a(y)\nabla_x u_0 + a(y)\nabla_y u_1) \\ &\quad = \nabla_y \cdot (a(y)\nabla_x u_0 + a(y)\nabla_y \tilde{u}_1 + a(y)\nabla_y (W(y) \cdot \nabla_x u_0(x))) \\ &\quad = \nabla_y \cdot (a(y)\nabla_x u_0 + a(y)\nabla_y \tilde{u}_1) \\ &\qquad + \nabla_y \cdot \left(a(y)\nabla_y \left(W_1(y)\frac{\partial u_0}{\partial x_1}(x) + W_2(y)\frac{\partial u_0}{\partial x_2}(x) \right) \right) \\ &\quad = \nabla_y \cdot (a(y)\nabla_x u_0 + a(y)\nabla_y \tilde{u}_1) \\ &\qquad + \nabla_y \cdot (a(y)\nabla_y W_1(y))\frac{\partial u_0}{\partial x_1}(x) + \nabla_y \cdot (a(y)\nabla_y W_2(y))\frac{\partial u_0}{\partial x_2}(x) \\ &\quad = \nabla_y \cdot (a(y)\nabla_x u_0 + a(y)\nabla_y \tilde{u}_1) \\ &\qquad - \nabla_y \cdot (a(y)e_1)\frac{\partial u_0}{\partial x_1}(x) - \nabla_y \cdot (a(y)e_2)\frac{\partial u_0}{\partial x_2}(x) \\ &\quad = \nabla_y \cdot (a(y)\nabla_x u_0 + a(y)\nabla_y \tilde{u}_1) - \nabla_y \cdot (a(y)\nabla_x u_0) \\ &\quad = \nabla_y \cdot (a(y)\nabla_y \tilde{u}_1), \end{aligned}$$

problem (2.86) is then equivalent to

$$\begin{cases} -\nabla_y \cdot (a(y)\nabla_y \tilde{u}_1) = 0, \\ \tilde{u}_1(x,\cdot) \text{ is } Y\text{-periodic for each given } x \in \Omega. \end{cases} \tag{2.87}$$

Problem (2.87) indicates that $\tilde{u}_1$ is independent of the choice of y, that is, $\tilde{u}_1(x, y) = \tilde{u}_1(x)$. Hence, it yields

$$u_1(x, y) = W(y) \cdot \nabla_x u_0(x) + \tilde{u}_1(x). \tag{2.88}$$

We rewrite problem (2.84) for u_2 (with respect to y):

$$\begin{cases} -\nabla_y \cdot (a(y)\nabla_y u_2) = -\nabla_y \cdot (q(y)u_1) - \nabla_x \cdot (q(y)u_0) \\ +\nabla_x \cdot (a(y)\nabla_x u_0) + \nabla_x \cdot (a(y)\nabla_y u_1) \\ +\nabla_y \cdot (a(y)\nabla_x u_1) + f(x), \\ \quad u_2(x, \cdot) \text{ is } Y\text{-periodic for each given } x \in \Omega. \end{cases} \tag{2.89}$$

By Lemma 7.1, the solvability condition assigned to (2.89) is as follows:

$$I_1 + I_2 + I_3 + I_4 + I_5 = 0,$$

where $I_1, \ldots, I_5$ are defined by

$$I_1 := -\int_Y (\nabla_y \cdot (q(y)u_1) + \nabla_x \cdot (q(y)u_0))\, dy,$$

$$I_2 := \int_Y \nabla_x \cdot (a(y)\nabla_x u_0) dy,$$

$$I_3 := \int_Y \nabla_x \cdot (a(y)\nabla_y u_1) dy,$$

$$I_4 := \int_Y \nabla_y \cdot (a(y)\nabla_x u_1) dy,$$

$$I_5 := \int_Y f(x) dy.$$

By (2.88) and (2.85), we have

$$\begin{aligned} I_1 &= -\int_Y ((\nabla_y \cdot q(y))u_1 + q(y) \cdot \nabla_y u_1 + q(y) \cdot \nabla_x u_0)\, dy \\ &= -\int_Y (q(y) \cdot \nabla_y (W(y) \cdot \nabla_x u_0(x) + \tilde{u}_1(x)) + q(y) \cdot \nabla_x u_0)\, dy \\ &= -\int_Y (q(y) \cdot \nabla_y (W(y) \cdot \nabla_x u_0(x)) + q(y) \cdot \nabla_x u_0)\, dy \\ &= -\sum_{i,j=1}^{2} \int_Y \left(q_j(y) \frac{\partial W_i}{\partial y_j}(y) \frac{\partial u_0}{\partial x_i} + q_i(y) \frac{\partial u_0}{\partial x_i} \right) dy \end{aligned}$$

$$\begin{aligned} I_1 &= -\sum_{i,j=1}^{2} \int_Y \left(q_j(y) \frac{\partial W_i}{\partial y_j}(y) + q_i(y) \right) dy \frac{\partial u_0}{\partial x_i} \\ &= -\left(\int_Y (I + \nabla_y W(y)) q(y) dy \right) \cdot \nabla_x u_0 \end{aligned}$$

We have

$$I_2 = \left(\int_Y a(y) dy \right) \nabla_x \cdot (\nabla_x u_0).$$

By (2.88), we have

$$\begin{aligned} I_3 &= \int_Y \nabla_x \cdot (a(y) \nabla_y (W(y) \cdot \nabla_x u_0(x) + \tilde{u}_1(x))) dy \\ &= \int_Y a(y) \nabla_x \cdot (\nabla_y (W(y) \cdot \nabla_x u_0(x))) dy \\ &= \sum_{i,j=1}^{2} \int_Y a(y) \frac{\partial}{\partial x_i} \left(\frac{\partial}{\partial y_i} \left(W_j(y) \frac{\partial u_0}{\partial x_j}(x) \right) \right) dy \\ &= \sum_{i,j=1}^{2} \int_Y a(y) \frac{\partial W_j}{\partial y_i}(y) \frac{\partial}{\partial x_i} \frac{\partial u_0}{\partial x_j}(x) dy \\ &= \sum_{i,j=1}^{2} \left(\int_Y a(y) \frac{\partial W_j}{\partial y_i}(y) dy \right) \frac{\partial}{\partial x_i} \frac{\partial u_0}{\partial x_j}(x) \\ &= \left(\int_Y a(y) \nabla_y W(y)^t dy \right) : \nabla_x \nabla_x u_0(x). \end{aligned}$$

Since the functions $a(y)$ and $u_1(x,y) = W(y) \cdot \nabla_x u_0(x) + \tilde{u}_1(x)$ are Y-periodic, by Gauß's divergence theorem,

$$I_4 = \int_Y \nabla_y \cdot (a(y) \nabla_x u_1) dy = \int_{\partial Y} a(y) (\nabla_x u_1) \cdot n(y) ds_y = 0.$$

We have

$$I_5 = |Y| f(x).$$

Therefore, we obtain the following macroscopic equation:

$$-\mathbb{D} : \nabla_x \nabla_x u_0(x) + \mathbb{Q} \cdot \nabla_x u_0(x) = f(x) \quad \text{for } x \in \Omega.$$

The corresponding effective diffusion tensor reads

$$\mathbb{D} := \frac{1}{|Y|}\int_Y a(y)(\mathbb{I} + \nabla_y W(y)^t)dy,$$

while the effective drift coefficient is

$$\mathbb{Q} := \frac{1}{|Y|}\int_Y (\mathbb{I} + \nabla_y W(y))q(y)dy.$$

The obtained macroscopic equation needs to be completed with homogeneous Dirichlet boundary conditions.

Solution 2.7.11. Regarding the indicated Exercises 1c and 2 from Holmes (1995), one can follow the procedure presented in this chapter. Essentially, one needs to redo the same type of calculations. Very few new terms will appear. The reader is invited to point them out. Another possible approach is to take the classical perturbation theory route as follows: The working procedure described in pp. 224–228 in Holmes (1995) can address the issue. However, due to the use of the explicit solutions for the auxiliary problems, the methodology is tailor-made for the one-dimensional case. Holmes's approach can be extended for slightly more complex situations, but it will be very difficult to use when perforations are included. Try it for fun, but make sure that you understand first how the first recommended strategy works.

Regarding the indicated Exercise 4 from *loc. cit.*, use the mean-value theorem for integrals.

Solution 2.7.12. Formulate the microscopic problem as a transmission problem by choosing $D_\varepsilon(x) = D_{1,\varepsilon}(x)$ if $x \in \Omega_\varepsilon^1$ and $D_\varepsilon(x) = D_{2,\varepsilon}(x)$ if $x \in \Omega_\varepsilon^2$, where Ω_ε^1 and Ω_ε^2 are the two periodic domains where the diffusion takes place.

An interesting related reading is Allaire (2012, Section 4.1), where the authors deal with the derivation of a double-porosity model.

Solution 2.7.13. Let ε be a small positive parameter. We make the following homogenization *ansatz*:

$$u_\varepsilon(x) = \sum_{i=0}^{\infty} \varepsilon^i u_i(x,y)\Big|_{y=\frac{x}{\varepsilon}}. \tag{2.90}$$

Inserting (2.90) into the target microscopic partial differential equation, we get

$$f_0(x) + \varepsilon^2 f_1(x,y) = \left(\nabla_x + \frac{1}{\varepsilon}\nabla_y\right) \cdot \left(-a(y)\left(\nabla_x + \frac{1}{\varepsilon}\nabla_y\right)\sum_{i=1}^{\infty}\varepsilon^i u_i(x,y)\right)$$
$$+ \partial_t \sum_{i=0}^{\infty}\varepsilon^i u_i(x,y) + b_\varepsilon \cdot \left(\nabla_x + \frac{1}{\varepsilon}\nabla_y\right)\sum_{i=0}^{\infty}\varepsilon^i u_i(x,y).$$

Expanding the sum, we have that

$$\begin{aligned}
&f_0(x) + \varepsilon^2 f_1(x,y) \\
&\quad = \partial_t u_0 + \varepsilon\partial_t u_1 + \varepsilon^2\partial_t u_2 + \text{h.o.t} + \left(\nabla_x + \frac{1}{\varepsilon}\nabla_y\right) \\
&\quad\quad \cdot \left(-\frac{1}{\varepsilon}a(y)\nabla_y u_0 + \varepsilon^0(-a(y)\nabla_x u_0 - a(y)\nabla_y u_1) + \varepsilon^1(-a(y)\nabla_x u_1\right. \\
&\quad\quad\quad \left. - a(y)\nabla_y u_2) + \text{h.o.t}\right) \\
&\quad\quad + b_\varepsilon \cdot (\varepsilon^{-1}\nabla_y u_0 + \varepsilon^0(\nabla_x u_0 + \nabla_y u_1) + \varepsilon^1(\nabla_x u_1 + \nabla_y u_2) + \text{h.o.t}).
\end{aligned}$$

We group together the terms with the same power of ε to obtain

$$\begin{aligned}
&\frac{1}{\varepsilon^2}\Big[\nabla_y \cdot (-a(y)\nabla_y u_0)\Big] \\
&\quad + \frac{1}{\varepsilon}\Big[\nabla_y \cdot (-a(y)\nabla_x u_0 - a(y)\nabla_y u_1) + \nabla_x \cdot (-a(y)\nabla_y u_0) + b_\varepsilon \cdot \nabla_y u_0\Big] \\
&\quad + \varepsilon^0\Big[\partial_t u_0 + \nabla_x \cdot (-a(y)\nabla_x u_0 - a(y)\nabla_y u_1) \\
&\quad\quad + \nabla_y \cdot (-a(y)\nabla_x u_1 - a(y)\nabla_y u_2) + b_\varepsilon \cdot (\nabla_x u_0 + \nabla_y u_1)\Big] \\
&\quad + \text{h.o.t} = f_0 + \varepsilon^2 f_1.
\end{aligned}$$

Proceeding now as when evaluating limits of indeterminate forms (as in L'Hopital's rule), we get the following set of equations that must hold as ε vanishes:

$$\begin{aligned}
T_{-2}[u_0] &:= \nabla_y \cdot (-a(y)\nabla_y u_0) = 0, \\
T_{-1}[u_0, u_1] &:= \nabla_y \cdot (-a(y)\nabla_x u_0 - a(y)\nabla_y u_1) \\
&\quad + \nabla_x \cdot (-a(y)\nabla_y u_0) + b(y) \cdot \nabla_y u_0 = 0,
\end{aligned}$$

$$\begin{aligned} T_0[u_0, u_1, u_2] := \partial_t u_0 &+ \nabla_x \cdot (-a(y)\nabla_x u_0 - a(y)\nabla_y u_1) \\ &+ \nabla_y \cdot (-a(y)\nabla_x u_1 - a(y)\nabla_y u_2) \\ &+ b(y) \cdot (\nabla_x u_0 + \nabla_y u_1) = f_0. \end{aligned}$$

We insert the *ansatz* (2.90) into the boundary conditions of the microscopic problem to guess the boundary conditions for equations (2.91). Upon expanding the Dirichlet boundary condition, we are led to

$$0 = \sum_{i=0}^{\infty} \varepsilon^i u_i(x, y) \quad \text{at } \Gamma^{ext} \quad \text{for each } \varepsilon > 0.$$

This implies that

$$u_0 = u_0(x) = 0 \quad \text{for } x \in \Gamma^{ext}.$$

We now consider the remaining boundary condition, i.e. the one posed on Γ_ε. Using the proposed *ansatz* gives

$$\begin{aligned} \varepsilon g_0 + \varepsilon^2 &= (-a_\varepsilon \nabla u_\varepsilon + b_\varepsilon u_\varepsilon) \cdot n_\varepsilon \\ &= \left(-a_\varepsilon \left(\nabla_x + \frac{1}{\varepsilon} \nabla_y \right) \sum_{i=0}^{\infty} \varepsilon^i u_i(x, y) + b_\varepsilon \sum_{i=0}^{\infty} \varepsilon^i u_i(x, y) \right) \cdot n_\varepsilon. \end{aligned}$$

Dividing both sides of the last identity by ε yields

$$g_0 + \varepsilon = \left(\sum_{i=0}^{\infty} \left[-a_\varepsilon \left(\varepsilon^{i-1} \nabla_x u_i + \varepsilon^{i-2} \nabla_y u_i \right) + \varepsilon^{i-1} b_\varepsilon u_i \right] \right) \cdot n_\varepsilon.$$

We now group the terms of the same order of ε to see that

$$\begin{aligned} &-\frac{1}{\varepsilon^2} a_\varepsilon \nabla_y u_0 \\ &-\frac{1}{\varepsilon} a_\varepsilon \nabla_x u_0 + \frac{1}{\varepsilon} b_\varepsilon u_0 - \frac{1}{\varepsilon} a_\varepsilon \nabla_y u_1 \\ &-\varepsilon^0 a_\varepsilon \nabla_x u_1 - \varepsilon^0 a_\varepsilon \nabla_y u_2 + \varepsilon^0 b_\varepsilon u_1 + \text{h.o.t} = g_0 + \text{h.o.t} \end{aligned}$$

This leads to the following relations on Γ_ε, holding as ε reaches zero:

$$\begin{cases} T_{-2}: & -a(y)\nabla_y u_0(x,y)\cdot n_\varepsilon = 0 \\ T_{-1}: & (-a(y)\nabla_x u_0(x,y) + b_\varepsilon u_0(x) - a(y)\nabla_y u_1(x,y))\cdot n_\varepsilon = 0 \\ T_0: & (-a(y)\nabla_x u_1(x,y) - a(y)\nabla_y u_2(x,y) + b_\varepsilon u_1(x,y))\cdot n_\varepsilon = g_0(x,y). \end{cases} \tag{2.91}$$

We consider the first auxiliary problem with the terms of order of ε^{-2}: Find $u_0 = u_0(x,y)$ such that for a.e. $x \in \Omega$, it holds that

$$\begin{cases} \nabla_y \cdot (-a(y)\nabla_y u_0(x,y)) = 0 & \text{in } Y, \\ -a(y)\nabla_y u_0(x,y)\cdot n(y) = 0 & \text{at } \partial Y_0, \end{cases} \tag{2.92}$$

where $u_0(x,\cdot)$ is Y-periodic. We impose the condition

$$\int_Y u_0(x,y)dy = c(x) \tag{2.93}$$

for some constant $c(x) \in \mathbb{R}$ to ensure the uniqueness of solutions to (2.92). We note here that the equation implies that $-a(y)\nabla_y u_0(x,y)$ must be constant in y. Therefore, the boundary condition implies that u_0 must be constant in y. So, we must have that $u_0 = u_0(x)$. We thus get from (2.93) that $c(x) = u_0(x)|Y|$.

The second auxiliary problem considers the terms of order of ε^{-1}: Find $u_1 = u_1(x,y)$ such that

$$\begin{cases} \nabla_y \cdot (-a(y)\nabla_x u_0 - a(y)\nabla_y u_1) \\ \qquad + \overbrace{\nabla_x \cdot (-a(y)\nabla_y u_0) + b(y)\cdot\nabla_y u_0}^{=0} = 0 \quad \text{in } Y, \\ (-a(y)\nabla_x u_0(x) + b(y)u_0(x) - a(y)\nabla_y u_1(x,y))\cdot n(y) = 0 \quad \text{at } \partial Y_0, \end{cases} \tag{2.94}$$

where $u_1(x,\cdot)$ is Y-periodic. According to Lemma 7.1, this problem has a unique weak solution (up to an additive constant) if and only if

$$\int_Y \nabla_y \cdot a(y)\nabla_x u_0(x)dy = \int_{\partial Y_0} (a(y)\nabla_x u_0(x) - b(y)u_0(x))\cdot n(y)d\sigma_y. \tag{2.95}$$

Applying Gauß's theorem on the left-hand side of (2.95) yields

$$\int_Y \nabla_y \cdot a(y)\nabla_x u_0(x)dy = \int_{\partial Y_0} a(y)\nabla_x u_0(x) \cdot n(y)d\sigma_y.$$

Also, since $\mathrm{div}(b) = 0$, we get that

$$0 = -\int_Y \nabla_y \cdot b(y)u_0(x)dy = \int_{\partial Y_0} -b(y)u_0(x) \cdot n(y)d\sigma_y.$$

Hence, we can see that (2.95) holds for problem (2.94); therefore, there exists a weak solution $u_1 = u_1(x, y)$ to (2.94) which is unique up to an additive constant.

We now introduce the cell functions $W(y) = (W_1(y), \ldots, W_d(y))^T \in \mathbb{R}^d$, where the functions W_j's are Y-periodic weak solutions to the cell problems

$$\begin{cases} -\nabla_y \cdot (a(y)\,(e_j + \nabla_y W_j(y))) = 0 \quad \text{in } Y, \\ a(y)\nabla_y W_j(y) \cdot n_j(y) = 0 \quad \text{at } \partial Y_0, \\ \displaystyle\int_Y W_j dy = 0. \end{cases} \tag{2.96}$$

We rewrite u_1 as

$$u_1(x, y) = W(y) \cdot \nabla_x u_0(x) + \tilde{u}_1(x). \tag{2.97}$$

Here, since we are considering a standard periodic homogenization scenario, we may set $\tilde{u}_1 = 0$.

Finally, we look at the last auxiliary problem, i.e. we collect the terms of the order of ε^0. The new problem reads as follows: Find $u_2 = u_2(x, y)$ such that

$$\begin{cases} \partial_t u_0 + \nabla_x \cdot (-a(y)\nabla_x u_0 - a(y)\nabla_y u_1) + \nabla_y \cdot (-a(y)\nabla_x u_1 - a(y)\nabla_y u_2), \\ \qquad + b(y) \cdot (\nabla_x u_0 + \nabla_y u_1) = f_0 \quad \text{in } Y, \\ (-a(y)\nabla_x u_1 - a(y)\nabla_y u_2 + b(y)u_1) \cdot n(y) = g_0 \quad \text{at } \partial Y_0, \end{cases} \tag{2.98}$$

where $u_2(x, \cdot)$ is Y-periodic. We note that the solvability condition must be satisfied. This implies

$$I_1 + I_2 + I_3 = I_4 + I_5,$$

where

$$I_1 := -\int_Y \partial_t u_0 dy + \int_Y \nabla_x \cdot (a(y)\nabla_x u_0(x))\, dy,$$

$$I_2 := \int_Y \nabla_x \cdot (a(y)\nabla_y u_1(x,y))\, dy + \int_Y \nabla_y \cdot (a(y)\nabla_x u_1(x,y))\, dy,$$

$$I_3 := \int_Y f_0(x) dy = |Y| f_0(x),$$

$$I_4 := \int_{\partial Y_0} a(y)\nabla_x u_1 \cdot n(y) d\sigma_y - \int_{\partial Y_0} b(y) u_1 \cdot n(y) d\sigma_y,$$

$$I_5 := \int_{\partial Y_0} g_0 d\sigma_y.$$

We have

$$I_1 = -|Y|\partial_t u_0(x) + \int_Y a(y) dy \nabla_x \cdot \nabla_x u_0(x).$$

For I_2 and I_4, we write $I_2 = I_{21} + I_{22}$ and $I_4 = I_{41} + I_{42}$. By Gauß's theorem, we see that $I_{22} = I_{41}$. Regarding the term I_{21}, we have

$$\begin{aligned} I_{21} &= \int_Y \nabla_x \cdot (a(y)\nabla_y u_1(x,y))\, dy \\ &= \int_Y a(y)\nabla_x \cdot \nabla_y (W(y) \cdot \nabla_x u_0(x)) dy \\ &= \int_Y a(y)\nabla_y W(y) dy : \nabla_x \nabla_x u_0(x). \end{aligned}$$

Looking now at I_{42}, we combine Gaußś theorem and (2.97) to obtain

$$\begin{aligned} I_{42} &= -\int_{\partial Y_0} (W(y) \cdot \nabla_x u_0(x))\, b(y) \cdot n(y) d\sigma_y \\ &= -\int_Y \nabla_y \cdot ((W(y) \cdot \nabla_x u_0(x)) b(y))\, dy \\ &= -\int_Y \nabla_y (W(y) \cdot b(y))\, dy \cdot \nabla_x u_0(x). \end{aligned}$$

Combining the above results and then dividing the outcome by $-|Z|$ yields the following macroscopic equation:

$$\frac{|Y|}{|Z|}\left(\partial_t u_0(x) - \fint_Y a(y)dy\nabla_x \cdot \nabla_x u_0(x) - \fint_Y a(y)\nabla_y W(y)dy : \nabla_x\nabla_x u_0(x)\right)$$

$$= \frac{|Y|}{|Z|}\left(f_0(x) + \int_Y \nabla_y (W(y) \cdot b(y))dy \cdot \nabla_x u_0(x)\right) - \frac{|\partial Y_0|}{|Z|}\fint_{\partial Y_0} g_0(x,y)d\sigma_y, \tag{2.99}$$

a.e. in Ω, with the boundary condition $u_0(x) = 0$ for all $x \in \partial\Omega$ and the initial condition $u_0(t = 0, x) = u_I^0(x) := \lim_{\varepsilon\to 0} u_I^\varepsilon(x)$.

Returning now to questions (Q1) and (Q2), we can say the following at this stage:

(i) We need to assume that the initial data converge to some bounded function $u_I^0 \in H^1(\Omega)$ as $\varepsilon \to 0$.

(ii) The assumption $\mathrm{div}(b_\varepsilon) = 0$ seems to be essential if one has in view the fact that equation (2.94) admits a unique solution if and only if

$$\int_{\partial Y_0} -b(y)u_0(x) \cdot n(y)d\sigma_y = 0,$$

which is a direct consequence of the fact that $\mathrm{div}(b_0) = 0$ in Y.

Solution 2.7.14. It makes sense to perform the two-scale asymptotic expansions for the case $\alpha = 1$. After performing the calculations, one obtains the insight required to see what other ranges in the parameter α may lead to potentially interesting limit results.

Solution 2.7.15. When solving this exercise, we will see that the two-scale asymptotics has a somewhat different structure here compared to what we expect to see in the standard Fickian diffusion case. We apply the following asymptotic expansions:

$$u_\varepsilon(t,x,y) = u_0(t,x,y) + \varepsilon u_1(t,x,y) + \varepsilon^2 u_2(t,x,y) + \cdots,$$

$$f_\varepsilon(t,x,y) = f_0(t,x) + \varepsilon^2 f_1(t,x,y),$$

with $y = \frac{x}{\varepsilon}$ and where the functions $f_1(x,\cdot)$ and $u_k(t,x,\cdot)$ $(k \in \mathbb{N})$ are Y-periodic for any given $(t,x) \in (0,\infty) \times \Omega$. We obtain

$$
\begin{aligned}
\partial_t u_\varepsilon &= \partial_t u_0 + \varepsilon \partial_t u_1 + \varepsilon^2 \partial_t u_2 \\
-\Delta(D_\varepsilon u_\varepsilon) &= -\nabla \cdot \nabla (D_\varepsilon u_\varepsilon) \\
&= -\left(\nabla_x + \frac{1}{\varepsilon}\nabla_y\right) \cdot \left(\nabla_x + \frac{1}{\varepsilon}\nabla_y\right)(D(y)(u_0 + \varepsilon u_1 + \varepsilon^2 u_2)) \\
&= -\Big[\varepsilon^{-2}\nabla_y \cdot \nabla_y (D(y)u_0) \\
&\quad + \varepsilon^{-1}(\nabla_x \cdot \nabla_y (D(y)u_0) + \nabla_y \cdot (\nabla_y (D(y)u_1) + \nabla_x (D(y)u_0))) \\
&\quad + \varepsilon^0(\nabla_x \cdot (\nabla_y (D(y)u_1) + \nabla_x (D(y)u_0)) \\
&\quad + \nabla_y \cdot (\nabla_y (D(y)u_2) + \nabla_x (D(y)u_1))) \\
&\quad + \varepsilon^1(\nabla_x \cdot (\nabla_y (D(y)u_2) + \nabla_x (D(y)u_1)) \\
&\quad + \nabla_y \cdot (\nabla_x (D(y)u_2))) + \varepsilon^2 \nabla_x \cdot \nabla_x (D(y)u_2)\Big]
\end{aligned}
$$

and

$$
\begin{aligned}
\varepsilon^\alpha \nabla u_\varepsilon \cdot b_\varepsilon &= \varepsilon^\alpha (\nabla_x + \frac{1}{\varepsilon}\nabla_y)(u_0 + \varepsilon u_1 + \varepsilon^2 u_2) \cdot b_\varepsilon \\
&= \varepsilon^\alpha \Big(\varepsilon^{-1}\nabla_y u_0 + \varepsilon^0(\nabla_x u0 + \nabla_y u_1) \\
&\quad + \varepsilon^1(\nabla_x u_1 + \nabla_y u_2) + \varepsilon^2 \nabla_x u_2\Big) \cdot b(y).
\end{aligned}
$$

As the next step, we expand the boundary conditions to obtain

$$
\begin{aligned}
&(-\nabla(D_\varepsilon u_\varepsilon) + \varepsilon^\alpha u_\varepsilon b_\varepsilon) \cdot n_\varepsilon \\
&\quad = \left(-\left(\nabla_x + \frac{1}{\varepsilon}\nabla_y\right)(D_\varepsilon u_\varepsilon) + \varepsilon^\alpha u_\varepsilon b_\varepsilon\right) \cdot n_\varepsilon \\
&\quad = \Bigg(-\left(\nabla_x + \frac{1}{\varepsilon}\nabla_y\right)(D(y)(u_0 + \varepsilon u_1 + \varepsilon^2 u_2)) \\
&\qquad + \varepsilon^\alpha (u_0 + \varepsilon u_1 + \varepsilon^2 u_2) b(y)\Bigg) \cdot n(y) \\
&\quad = \Big(-\varepsilon^{-1}\nabla_y (D(y)u_0) - \varepsilon^0(\nabla_x (D(y)u_0) + \nabla_y (D(y)u_1)) \\
&\qquad - \varepsilon^1(\nabla_x (D(y)u_1) + \nabla_y (D(y)u_2)) - \varepsilon^2 \nabla_x (D(y)u_2)\Big) \cdot n(y) \\
&\qquad + \varepsilon^\alpha (u_0 + \varepsilon u_1 + \varepsilon^2 u_2) b(y) \cdot n(y).
\end{aligned}
$$

By collecting all possible terms containing as factor ε^{-2}, we are led to the expressions

$$\begin{cases} -\nabla_y \cdot \nabla_y \big(D(y)u_0\big) + \varepsilon^\alpha \Big(\varepsilon^{-1}\nabla_y u_0 + \varepsilon^0(\nabla_x u0 + \nabla_y u_1) \\ \qquad + \varepsilon^1(\nabla_x u_1 + \nabla_y u_2) + \varepsilon^2 \nabla_x u_2\Big) \cdot b(y) = 0 \quad \text{in } Y, \\ \varepsilon^{-1}(\nabla_y \big(D(y)u_0\big) \cdot n(y) + \varepsilon^\alpha (u_0 + \varepsilon u_1 + \varepsilon^2 u_2) b(y) \cdot n(y) \\ \qquad = \varepsilon^\beta(\partial_t u_0 + \varepsilon \partial_t u_1 + \varepsilon^2 \partial_t u_2) \quad \text{on } \Gamma. \end{cases}$$

If we want to obtain simple equations, then the natural assumption would be $\alpha > -1$. However, this intuition fails us, as this choice leads to significant information loss. Instead, we make the choice of $\alpha = -1$. This retains some terms depending on the divergence-free velocity vector $b(y)$. We also choose $\beta > -1$. With these choices at hand, we can now formulate our first auxiliary problem: Find u_0 such that

$$\begin{cases} -\nabla_y \cdot \nabla_y \big(D(y)u_0\big) + \nabla_y u_0 b(y) = 0 \quad \text{in } Y, \\ \big(-\nabla_y(D(y)u_0) + u_0 b(y)\big) \cdot n(y) = 0 \quad \text{on } \Gamma \end{cases}$$

Assuming

$$u_0(t, x, y) = W(y)\tilde{u}_0(t, x),$$

we may obtain a boundary-value problem to determine $W(y)$. This can be seen as a cell problem: Find w such that

$$\begin{cases} -\nabla_y \cdot \nabla_y \big(D(y)W(y)\big) + \nabla_y W(y) b(y) = 0 \quad \text{in } Y, \\ \big(-\nabla_y(D(y)W(y) + W(y))b(y)\big) \cdot n(y) = 0 \quad \text{on } \Gamma, \\ W(\cdot) \text{ is Y-periodic.} \end{cases}$$

Collecting the remaining terms containing ε^{-1} and using what we know so far,

$$\begin{cases} -\nabla_y \cdot \nabla_y(D(y)u_1) + \nabla_y u_1 b(y) = \nabla_x \cdot \nabla_y(D(y)u_0) \\ \qquad + \nabla_y \cdot \nabla_x(D(y)u_0) - \nabla_x u_0 b(y) \quad \text{for all } y \in Y, \\ \nabla_y\big(D(y)u_1\big)n(y) + u_1 b(y) \cdot n(y) = \nabla_x\big(D(y)u_0\big) \cdot n(y) \\ \qquad + \varepsilon^\beta(\partial_t u_0 + \varepsilon\partial_t u_1 + \varepsilon^2 \partial_t u_2) \quad \text{for all } y \in \Gamma. \end{cases}$$

We pick $\beta > 0$. For the compatibility condition to be fulfilled, we need to show that

$$\int_Y [\nabla_x \cdot \nabla_y (D(y)u_0) + \nabla_y \cdot \nabla_x (D(y)u_0) - \nabla_x u_0 b(y)]\, dy \quad (2.100)$$

$$= \int_\Gamma [\nabla_x (D(y)u_0) \cdot n(y)]\, d\sigma_y. \quad (2.101)$$

The second term on the left-hand side of (2.101) balances the corresponding term on the right-hand side by the application of Gauß's theorem. Then, we have

$$\int_Y (\nabla_x \cdot \nabla_y (D(y)u_0) - \nabla_x u_0 b(y))\, dy = 0. \quad (2.102)$$

We now insert the *ansatz* for u_0, that is, $u_0(t,x,y) = W(y)\tilde{u}_0(t,x)$, to be able to write

$$\int_Y (\nabla_x \tilde{u}_0(t,x) \nabla_y (D(y)W(y)) - W(y)b(y)\nabla_x \tilde{u}_0(t,x))\, dy = 0. \quad (2.103)$$

The expression $\nabla_x \tilde{u}_0(t,x)$, which appears in both terms, is unaffected by the integral and can be divided out. Doing this yields

$$\int_Y [\nabla_y (D(y)W(y)) - W(y)b(y)]\, dy = 0. \quad (2.104)$$

This expression has a somewhat similar structure as our cell problem. If we take the cell problem and multiply it by $y \in Y$ and, finally, integrate the result by parts on Y, then we receive the above expression together with a boundary term. We have

$$\int_Y \nabla_y [-\nabla_y (D(y)W(y)) + W(y)b(y)] y dy$$

$$= \int_Y \nabla_y (D(y)W(y)) - W(y)b(y) dy$$

$$- \int_\Gamma (\nabla_y (D(y)W(y)) - W(y)b(y)) y \cdot n(y) dy = 0.$$

Taking a closer look at the second boundary term, we demand that the normal flow of $b(y)$ around Γ must vanish. The first term of the boundary integral is zero due to the fundamental theorem of calculus and the periodicity of $D(y)W(y)$. Since both boundary terms vanish, we have that the required compatibility condition is satisfied.

We restate the problem with all we have learned and assumed:
Find u_1 such that

$$\begin{cases} -\nabla_y \cdot \nabla_y(D(y)u_1) + \nabla_y u_1 b(y) = \nabla_x \cdot \nabla_y(D(y)u_0) \\ \qquad + \nabla_y \cdot \nabla_x(D(y)u_0) - \nabla_x u_0 b(y) \quad \text{for all } y \in Y, \\ \nabla_y(D(y)u_1)n(y) = \nabla_x(D(y)u_0) \cdot n(y) \quad \text{for all } y \in \Gamma. \end{cases}$$

We proceed as before by collecting all the terms with ε^0. Consequently, we obtain the following problem: Find u_2 such that

$$\begin{cases} -\nabla_y \cdot \nabla_y(D(y)u_2) + \nabla_y u_2 \\ \qquad = f_0(x) - \partial_t u_0 + \nabla_x \cdot (\nabla_y(D(y)u_1) + \nabla_x(D(y)u_0)) \\ \qquad\quad + \nabla_y \cdot \nabla_x(D(y)u_1)) - \nabla_x u_1 \\ (-\nabla_y(D(y)u_2) + u_2 b(y)) \cdot n(y) = \nabla_x(D(y)u_1) \cdot n(y). \end{cases}$$

At a first glance, it might seem as though we will not obtain a closed system from this auxiliary problem. But if we look at the compatibility condition for the above problem, we see that it is in fact possible to find a equation for $\tilde{u}_0(x)$. To show this, we start by using our *ansatz* for u_0, together with yet another *ansatz* for u_1, i.e. $u_1(x,y) = \chi(y)\nabla_x \tilde{u}_0(x)$. We start with the part of the equation that is posed within the domain:

$$\begin{aligned} f_0 - \partial_t u_0 &+ \nabla_x \cdot (\nabla_y(D(y)u_1) + \nabla_x(D(y)u_0)) + \nabla_y \cdot \nabla_x(D(y)u_1)) - \nabla_x u_1 \\ &= f_0 - W(y)\partial_t \tilde{u}_0 + \Delta_x \tilde{u}_0 \nabla_y(D(y)\chi(y)) + D(y)W(y)\Delta_x \tilde{u}_0 \\ &\quad + \nabla_y(D(y)\chi(y))\Delta_x \tilde{u}_0 - \chi(y)\Delta_x \tilde{u}_0. \end{aligned}$$

If we apply the same substitutions to the boundary term inside the compatibility condition, then we obtain the next term:

$$\nabla_x(D(y)u_1) \cdot n(y) = D(y)\chi(y)\Delta_x \tilde{u}_0 \cdot n(y).$$

We observe that if we integrate the result over ∂Y and apply Gauß's theorem, we obtain a term that balances half of a term arising in the previous expression. If we cancel the right-hand side and reorder the terms, then we obtain the wanted macroscopic equation in terms of the unknown $\tilde{u}_0$, together with the corresponding effective coefficients, viz. for all $t > 0$ and

$x \in \Omega$, find the function $\tilde{u}_0 = \tilde{u}_0(t, x)$ satisfying

$$\partial_t(a_{\text{eff}}\tilde{u}_0) + \Delta_x(-\mathbb{D}_{\text{eff}}\tilde{u}_0) = f_0 \quad \text{in } \Omega,$$

where the effective coefficients a_{eff} and $\mathbb{D}_{\text{eff}}$ are defined via

$$a_{\text{eff}} := \frac{1}{|Y|}\int_Y W(y)dy,$$

and

$$\mathbb{D}_{\text{eff}} := \frac{1}{|Y|}\int_Y (\nabla_y(D(y)\chi(y)) + D(y)W(y) - \chi(y))dy.$$

At this point, we note that the obtained macroscopic problem becomes computable. To see this, recall that we already have a boundary-value problem for solving $W(\cdot)$, while the behavior of $\tilde{u}_0$ across the boundary $\partial\Omega$ is governed by an homogeneous Dirichlet condition. If, additionally, we perform the corresponding substitutions in the auxiliary problem for u_1, then we obtain in a straightforward way an elliptic boundary-value problem determining uniquely the second cell function $\chi(\cdot)$. It will turn out that $\chi(\cdot)$ depends on the drift $b(\cdot)$. Consequently, the tensor $\mathbb{D}_{\text{eff}}$ will depend naturally on both local diffusion and local drift effects, which makes it a true dispersion tensor in terms of standard porous-media terminology; cf. Bear (1988).

It is worth noting that this exercise is in fact a complex one and the provided solution is not exhaustive. Further scaling options (i.e. other selections of ranges for the parameters α and β) can still be explored. We point the reader to a very recent paper by Amar *et al.* (2025) on precisely this topic.

Chapter 3

Homogenization of a Linear Second-Order Elliptic Equation: A Two-Scale Convergence Approach

The aim of this chapter is to bring the concept of two-scale convergence (as introduced in Definition 7.3) to "action" in a swift and efficient manner. It turns out that such a convergence tool mimics well the two-scale asymptotics ideas introduced in Chapter 2. In fact, it discovers the leading terms in the homogenization *ansatz* as proposed in the two-scale asymptotics expansion (2.37). Two-scale convergence arguments are used here to perform the homogenization of a linear second-order elliptic equation posed in a fixed (non-oscillatory) domain Ω. This scenario sheds light on many of the standard but inherent technicalities one needs to face when performing rigorously an homogenization exercise.[1] The way this chapter is structured is very much inspired by the lecture notes by G. Allaire (2002a). We advise the reader to study the well-written methodological paper by Lukkassen *et al.* (2002), with a special focus on Section 4 therein.

3.1 Setting of the Microscopic Problem: Working Hypotheses

Relying on the assumed two-scale expansion of the oscillating microscopic solution, we were able to derive in the previous chapter various upscaled

[1]The reader should ideally be familiar with the content of Chapter 7. In particular, the information presented in Section 7.6 is essential for following the arguments in this chapter.

equations and effective coefficients. This is an efficient procedure to quickly get results, but it is a rather formal way of proceeding, as we do not know for sure if such a two-scale structure of the unknown solution holds. To get the needed trust in the formal asymptotic calculations, we wish to learn how to perform rigorously[2] the passage to the homogenization limit $\varepsilon \to \infty$. To illustrate the main ideas, we work with the easiest linear second-order elliptic equation.

Find $u_\varepsilon \in H_0^1(\Omega)$ such that

$$\operatorname{div}\big(-a_\varepsilon(x)\nabla u_\varepsilon(x)\big) = f(x) \quad \text{for } x \in \Omega \subset \mathbb{R}^d, \tag{3.1}$$

$$u_\varepsilon(x) = 0 \quad \text{for } x \in \partial\Omega, \tag{3.2}$$

where $a_\varepsilon(x) = A(\frac{x}{\varepsilon}) \in \mathbb{R}^{d^2}$.

Equations (3.1)–(3.2) are collected as problem (P_ε). We also refer to (P_ε) as the microscopic problem.

We consider the following assumptions to be fulfilled:

(H1) The matrix A satisfies the following coercivity condition: There exist $\alpha, \beta \in (0,\infty)$ independent of ε such that

$$\alpha||\zeta||^2_{\mathbb{R}^d} \leq \sum_{i,j=1}^{d} A_{ij}\left(\frac{x}{\varepsilon}\right)\zeta_i\zeta_j \leq \beta||\zeta||^2_{\mathbb{R}^d} \tag{3.3}$$

holds for a.e. $x \in \Omega$ and all $\zeta := (\zeta_1, \ldots, \zeta_d) \in \mathbb{R}^d$.

(H2) $A_{ij} \in L^\infty_\#(Y)$ for all $(i,j) \in \{1,\ldots,d\}^2$, $Y \subset \Omega$ a periodic patch, A symmetric.

(H3) $f \in L^2(\Omega)$.

(H4) Ω is a $\mathbb{R}^d$-parallelepiped, while Y is a hypercube of unit volume, i.e. $|Y| = 1$.

Assumptions (H1)–(H4) can be relaxed. For instance, the matrix A does not have to be *per se* symmetric; instead of (H3), one could have as well $f \in H^{-1}(\Omega)$; or Ω could have curved boundaries. However, in this chapter, we restrict our attention to (H1)–(H4), simply because these assumptions

[2]The word "rigorous" refers here to the fact that we will not assume anymore *a priori* that the two-scale expansion holds.

offer the simplest setting where the technique of two-scale convergence is applicable. As a rule of thumb, if not otherwise mentioned, the surface $\partial\Omega$ is assumed to be Lipschitz.

3.1.1 *Homogenization procedure*

As standing hypotheses, we assume (H1)–(H4). For methodological reasons, we split the so-called homogenization procedure into six conceptually different steps. We proceed as follows:

Step 1: Well-posedness of the microscopic problem. The first thing to do is to define a good concept of weak solution to the microscopic problem (P_ε). Then, we need to ensure its well-posedness.

In this spirit, we propose the following concept of solution to the model problem (P_ε).

Definition 3.1 (Weak solution to (P_ε)). We refer to $u_\varepsilon \in H_0^1(\Omega)$ as a weak solution to (P_ε) if and only if, for all $\varphi \in H_0^1(\Omega)$, the following identity holds:

$$(a_\varepsilon \nabla u_\varepsilon, \nabla \varphi) = (f, \varphi). \tag{3.4}$$

A straightforward application of the Lax–Milgram lemma (cf. Theorem 7.1; see Ciorănescu and Donato (1999, Theorem 4.6) as well) provides the existence and uniqueness of $u_\varepsilon \in H_0^1(\Omega)$ solving (P_ε) in the sense of Definition 3.1. Showing the stability of u_ε with respect to the data of (P_ε) is an easy-to-do exercise; we leave this as homework.

Step 2: ε-independent *a priori* estimates. Any rigorous asymptotic study needs parameter-independent upper bounds. Getting such bounds is precisely what we aim here. Choosing in (3.4) the test function $\varphi = u_\varepsilon$ and using the coercivity condition (H1), we obtain

$$\begin{aligned} \alpha \|u_\varepsilon\|^2_{H_0^1(\Omega)} &\le (a_\varepsilon \nabla u_\varepsilon, \nabla u_\varepsilon) = (f, u_\varepsilon) \le \|f\|_{L^2(\Omega)} \|u_\varepsilon\|_{L^2(\Omega)} \\ &\le c_P \|f\|_{L^2(\Omega)} \|u_\varepsilon\|_{H_0^1(\Omega)}, \end{aligned}$$

where the constant $c_P > 0$ is from Poincaré's inequality[3] (recall that $H_0^1(\Omega) \hookrightarrow L^2(\Omega)$). Dividing the latter expression by $\|u_\varepsilon\|_{H_0^1(\Omega)}$, which is

[3] The reader may wonder whether there is a different way to derive an upper bound on $\|u_\varepsilon\|_{H_0^1(\Omega)}$ without using Poincaré's inequality.

genuinely a non-vanishing quantity when f is not identically zero, we are led to the estimate

$$\|u_\varepsilon\|_{H_0^1(\Omega)} \leq \frac{c_P}{\alpha}\|f\|_{L^2(\Omega)}. \tag{3.5}$$

This inequality is often named the "energy estimate". Note that the constant $\frac{c_P}{\alpha}$ from inequality (3.5) is independent of ε.

Step 3: Compactness arguments preparing the passage $\varepsilon \to 0$. This part is essentially the true homogenization step, as the macroscopic limit u_0 is unveiled here. Now, we note that the hypotheses of the two-scale compactness theorem (see Theorem 7.7) are fulfilled. Hence, we can apply these compactness arguments to our setting.

Since by (3.5) the sequence of functions (u_ε) is uniformly bounded in $H_0^1(\Omega)$, there exist limit functions u_0 and u_1[4] such that

$$u_0 \in L^2(\Omega \times Y),$$
$$u_1 \in L^2(\Omega; H^1_\#(Y)/\mathbb{R}),$$

satisfy the convergences

$$u_\epsilon \overset{2}{\rightharpoonup} u_0,$$
$$\nabla u_\epsilon \overset{2}{\rightharpoonup} \nabla_x u_0 + \nabla_y u_1.$$

On the other hand, from the fact that the sequence (u_ε) is uniformly bounded in $H_0^1(\Omega)$ and from the compactness of the embedding $H_0^1(\Omega) \hookrightarrow L^2(\Omega)$ (cf. e.g. Theorem 7.5), we deduce that (up to subsequences) $u^\epsilon \rightharpoonup \hat{u}_0(x,y)$ (weakly), and respectively, $u_\varepsilon \to \bar{u}_0(x)$ (strongly). Consequently, by the uniqueness of the weak limit, jointly with arguments from Theorem 5 in Lukkassen *et al.* (2002), we are led to

$$\bar{u}_0(x) = \hat{u}_0(x,y) = u_0(x) \quad \text{for a.e. in } (x,y) \in \Omega \times Y. \tag{3.6}$$

Essentially, (3.6) indicates that the limit function u_0 is independent of the microscopic variable y.

Step 4: Weak formulation of the two-scale limit problem. Set $x \in \Omega$ arbitrarily. Choosing in Definition 3.1 as the test function the expression

$$\varphi(x) = \phi_0(x) + \varepsilon\phi_1\left(x, \frac{x}{\varepsilon}\right), \tag{3.7}$$

[4]We expect already at this stage that these functions u_0 and u_1 are the first two terms from the homogenization *ansatz* (2.37).

with

$$(\phi_0, \phi_1) \in C_0^\infty(\Omega) \times C_0^\infty(\Omega; C_\#^\infty(Y)),$$

we obtain

$$\int_\Omega A\Big(\frac{x}{\varepsilon}\Big)\nabla u_\varepsilon(x)\nabla\varphi(x)dx = \int_\Omega f(x)\varphi(x)dx.$$

Hence, it holds

$$\int_\Omega A\Big(\frac{x}{\varepsilon}\Big)\nabla u_\varepsilon\Big(\nabla_x\phi_0(x) + \nabla_y\phi_1\Big(x, \frac{x}{\varepsilon}\Big) + \varepsilon\nabla_x\phi_1\Big(x, \frac{x}{\varepsilon}\Big)\Big)dx$$
$$= \int_\Omega f(x)\phi_0(x)dx + \varepsilon\int_\Omega f(x)\phi_1\Big(x, \frac{x}{\varepsilon}\Big)dx.$$

Rearranging the resulting terms, we have

$$\int_\Omega A\Big(\frac{x}{\varepsilon}\Big)\nabla u_\varepsilon\Big(\nabla_x\phi_0(x) + \nabla_y\phi_1\Big(x, \frac{x}{\varepsilon}\Big)\Big) + \varepsilon\int_\Omega A\Big(\frac{x}{\varepsilon}\Big)\nabla u_\varepsilon\nabla_x\phi_1\Big(x, \frac{x}{\varepsilon}\Big)dx$$
$$= \int_\Omega f(x)\phi_0(x)dx + \varepsilon\int_\Omega f(x)\phi_1\Big(x, \frac{x}{\varepsilon}\Big)dx.$$

Passing now with $\varepsilon \to 0$ in each of the terms of the previous identity, we deduce the weak form of the limit two-scale problem. To this end, we now treat separately each of the following terms[5]:

$$I_1 := \int_\Omega A\Big(\frac{x}{\varepsilon}\Big)\nabla u_\varepsilon\nabla\phi_0(x)dx,$$

$$I_2 := \int_\Omega A\Big(\frac{x}{\varepsilon}\Big)\nabla u_\varepsilon\nabla_y\phi_1\Big(x, \frac{x}{\varepsilon}\Big)dx,$$

$$I_3 := \varepsilon\int_\Omega A\Big(\frac{x}{\varepsilon}\Big)\nabla u_\varepsilon\nabla_x\phi_1\Big(x, \frac{x}{\varepsilon}\Big)dx,$$

$$I_4 := \int_\Omega f(x)\phi_0(x)dx,$$

$$I_5 := \varepsilon\int_\Omega f(x)\phi_1\Big(x, \frac{x}{\varepsilon}\Big)dx.$$

[5] A methodological hint: Since this is for the first time students encounter a concrete application of the two-scale convergence, they may be asked to perform in front of the class these five passages to the two-scale limit. In that case, the terms should be chosen from simple (i.e. I_4) to complicated (i.e. $I_1 + I_2$).

Firstly, we note that $\lim_{\varepsilon\to 0} I_4 = \int_\Omega f(x)\phi(x)dx$ as the right-hand side of I_4 is independent of ε. Since $f \in L^2(\Omega)$ and $\phi_1 \in L^2(\Omega \times Y)$, it yields $f(x)\phi_1\left(x, \frac{x}{\varepsilon}\right) \in L^1(\Omega)$ uniformly in ε. Hence, we have $\lim_{\varepsilon\to 0} I_5 = 0$. A similar argument ensures that $\lim_{\varepsilon\to 0} I_3 = 0$. What concerns the last two terms, it is convenient to treat them together as follows:

$$\begin{aligned}\lim_{\varepsilon\to 0}(I_1 + I_2) &= \lim_{\varepsilon\to 0}\int_\Omega A\Big(\frac{x}{\varepsilon}\Big)\nabla u_\varepsilon\Big(\nabla_x\phi_0(x) + \nabla_y\phi_1\Big(x, \frac{x}{\varepsilon}\Big)\Big)dx \\ &= \lim_{\varepsilon\to 0}\int_\Omega \nabla u_\varepsilon A^t\Big(\frac{x}{\varepsilon}\Big)\Big(\nabla_x\phi_0(x) + \nabla_y\phi_1\Big(x, \frac{x}{\varepsilon}\Big)\Big)dx \\ &= \int_{\Omega\times Y} (\nabla_x u_0(x) + \nabla u_1(x,y))\,A^t(y) \\ &\qquad \times (\nabla\phi_0(x) + \nabla_y\phi_1(x,y))\,dxdy.\end{aligned}$$

To obtain the last equality, we have used Theorem 7.7 to describe what happens with ∇u_ε under two-scale convergence, and we employed as a test function the expression

$$\Phi\Big(x, \frac{x}{\varepsilon}\Big) := A^t\Big(\frac{x}{\varepsilon}\Big)\Big(\nabla_x\phi_0(x) + \nabla_y\phi_1\Big(x, \frac{x}{\varepsilon}\Big)\Big).$$

We are allowed to do so, as $\Phi \in L^2_\#(Y, C(\Omega))$; hence, it is admissible as a test function. Finally, we obtain

$$\begin{aligned}&\int_\Omega\int_Y (\nabla_x u_0(x) + \nabla_y u_1(x,y))\,A^t(y)(\nabla_x\phi_0(x) + \nabla_y\phi_1(x,y))\,dxdy \\ &\qquad = \int_\Omega\int_Y f(x)\phi_0(x)dxdy,\end{aligned}$$

and hence, since $f = f(x)$ and $|Y| = 1$, it holds that

$$\begin{aligned}&\int_\Omega\int_Y A(y)(\nabla_x u_0(x) + \nabla_y u_1(x,y))\,(\nabla_x\phi_0(x) + \nabla_y\phi_1(x,y))\,dxdy \\ &\qquad = \int_\Omega f(x)\phi_0(x)dx. \qquad (3.8)\end{aligned}$$

When looking at (3.8), it is worthwhile to note that two length scales are prominent: One length scale involves the macroscopic variable $x \in \Omega$, while the other is linked to the microscopic variable $y \in Y$. These length scales are "separated" from each other. We use the identity in (3.8) to define the weak formulation of the limit two-scale problem. Let us call this problem (P_0). It reads as follows:

Find the pair

$$(u_0, u_1) \in H_0^1(\Omega) \times L^2(\Omega; H_\#^1(Y)/\mathbb{R})$$

such that the identity

$$\int_{\Omega\times Y} A(y)(\nabla_x u_0 + \nabla_y u_1(x,y))(\nabla_x \phi_0 + \nabla_y \phi_1(x,y))dxdy = \int_\Omega f\phi_0 dx$$

holds for all pairs of test functions

$$(\phi_0, \phi_1) \in H_0^1(\Omega) \times L^2(\Omega; H_\#(Y)/\mathbb{R}).$$

This two-scale formulation is, in fact, the main result of the asymptotics work.

Step 5: Weak solvability of the limit two-scale problem. It is, of course, essential that the resulting two-scale problem (P_0) is well-posed. This fact is not *a priori* clear and requires a proof. We are particularly concerned with the existence and uniqueness of weak solutions. As the problem is linear, the stability of solutions with respect to data and parameters follows if the first two mentioned properties hold.

The limiting process $\varepsilon \to 0$ finds (by means of a subsequence) the weak solution to (P_0). The procedure to pass by two-scale convergence to the limit, guaranteed by two-scale compactness arguments, offers an elegant proof for the existence[6] of solutions to the limit two-scale problem. Furthermore, since the limit object is linear, one can easily show by standard

[6] Alternatively, one can apply the Lax–Milgram lemma in the context of the Hilbert space

$$\mathcal{H} := H_0^1(\Omega) \times L^2(\Omega; H_\#^1(Y)/\mathbb{R})$$

endowed with the norm

$$\|(u_0, u_1)\|_{\mathcal{H}} := \|\nabla_x u_0\|_{L^2(\Omega)} + \|\nabla_y u_1\|_{L^2(\Omega\times Y)}.$$

partial differential equations arguments[7] that the uniqueness of weak solutions to (P_0) also holds true.

Step 6: Strong formulation of the two-scale limit problem. One may say that the work of the applied mathematician asked to average (homogenize or upscale) the microscopic problem (P_ε) is concluded once the structure of the limit two-scale problem (P_0) is derived. However, despite the fact that the obtained structure is precisely what one would need for suggesting a finite element approximation of the corresponding solution, the result (3.8) is rather hard to explain to the applied scientist, owner of the problem (P_ε) formulated for some engineering scenario. The role of this step is to emphasize the strong formulation of the two-scale limit problem, from where the effective coefficients can be easily read off.

To proceed in this direction, we rewrite (P_0) to obtain

$$\begin{aligned}\int_\Omega\int_Y A(y)\nabla_x u_0(x)\nabla_x\phi_0(x)dxdy + \int_\Omega\int_Y A(y)\nabla_y u_1(x,y)\nabla_x\phi_0(x)dxdy\\ + \int_\Omega\int_Y A(y)\nabla_x u_0(x)\nabla_y\phi_1(x,y)dxdy\\ + \int_\Omega\int_Y A(y)\nabla_y u_1(x,y)\nabla_y\phi_1(x,y)dxdy\\ = \int_\Omega f(x)\phi_0(x)dx.\end{aligned}$$

This leads us to

$$\begin{aligned}\int_\Omega -\mathrm{div}_x\left(\int_Y A(y)dy(\nabla_x u_0(x) + \nabla_y u_1(x,y))\right)\phi_0(x)dx - \int_\Omega f(x)\phi_0(x)dx\\ = \int_{\Omega\times Y}\mathrm{div}_y\Big(A(y)(\nabla_x u_0(x) + \nabla_y u_1(x,y))\Big)\phi_1(x,y)dxdy\end{aligned}$$

for all $(\phi_0,\phi_1)\in C_0^\infty(\Omega)\times C_0^\infty(\Omega;C_\#^\infty(Y))$. Choosing first $\phi_0 = 0$ and, subsequently, $\phi_1 = 0$ within the domains Ω and respectively within $\Omega\times Y$, we obtain the strong formulation of the two-scale limit problem after applying twice the fundamental lemma of the calculus of variations (or the so-called Du Bois–Reymond lemma).

[7]The uniqueness of weak solutions can be obtained either directly via the application of the Lax–Milgram lemma or by an energy-type estimate involving testing conveniently with the difference of two distinct weak solutions to (P_ε) with precisely the difference in these solutions.

Find the pair of functions

$$(u_0, u_1) \in H_0^1(\Omega) \times L^2(\Omega; H^1_{\#}(Y)/\mathbb{R})$$

satisfying weakly Equations (3.9)–(3.11):

$$-\mathrm{div}_y\Big(A(y)(\nabla_x u_0(x) + \nabla_y u_1(x,y))\Big) = 0 \quad \text{for a.e. } (x,y) \in \Omega \times Y, \quad (3.9)$$

$$-\mathrm{div}\left(\int_Y A(y)(\nabla_x u_0(x) + \nabla_y u_1(x,y))\right) = f(x) \quad \text{for a.e. } x \in \Omega, \quad (3.10)$$

$$u_1 = 0 \quad \text{on } \partial\Omega \quad \text{and} \quad u_1 \text{ is } Y\text{-periodic.} \quad (3.11)$$

Note that (3.9) is precisely the structure seen for the auxiliary problem T_{-2} in the formal homogenization procedure, as explained in Section 2.3 (which appeared as a result of collecting the terms of order of ε^{-2}). The role of this auxiliary problem was to introduce cell problems that can eliminate the term u_1 in the expansion of u_ε. In the same spirit, we can use the boundary-value problem (3.9) to eliminate u_1 from the strong (and weak) formulation of the limit two-scale problem.

As we have seen earlier, the solution to the cell problem admits the representation

$$u_1(x,y) = -\sum_{j=1}^{d} W_j(y)\partial_{x_j} u_0 + \tilde{u}_1(x), \quad (3.12)$$

for some function $\tilde{u}_1$, where the cell function $W = (W_1, \ldots, W_d) \in H^1_{\#}(Y)$ is a (weak) solution to the cell problem

$$-\mathrm{div}_y(A(y)\nabla_y W_j(y)) = -\sum_{i=1}^{d} \partial_{y_i} A_{ij}(y) \quad \text{for } y \in Y,$$

$$\int_Y W_j(y)dy = 0 \quad \text{and} \quad W_j(\cdot) \text{ is } Y\text{-periodic for all } j \in \{1, \ldots, d\}.$$

Substituting now u_1 defined by (3.12) into (3.10), we obtain the desired upscaled (or homogenized or macroscopic) partial differential equation

$$-\sum_{i,k=1}^{d}\left[\sum_{j=1}^{d}\int_Y (A_{ik} + A_{ij}\partial_{y_j} W_k(y))\, dy\right] \partial^2_{x_i x_k} u_0 = f \quad \text{in } \Omega \quad (3.13)$$

with the homogeneous Dirichlet boundary condition posed across $\partial\Omega$. Note that we have used repeatedly the fact that $|Y| = 1$.

Looking now at the structure of (3.13), we introduce the tensor $\mathbb{D}$, defined by

$$D_{ik} := \sum_{j=1}^{d} \int_Y \left(A_{ik} + A_{ij}\partial_{y_j} W_k(y)\right) dy. \tag{3.14}$$

Here, $\mathbb{D} = [D_{ik}]$ is the so-called *effective* (macroscopic) diffusion tensor and enters as a coefficient in the upscaled model equation.

It is worth noting that if at the end of Step 4 one makes use of the representation of u_1 via (3.12), then Step 5 becomes much easier, as the weak form of two-scale formulation simplifies considerably. This approach works well when the formal homogenization asymptotics for the given microscopic problem has already been performed, as it ensures that the structure of the cell functions is already known.

3.1.2 *Two key properties of the effective diffusion tensor*

We show now the symmetry and ellipticity of the tensor $\mathbb{D}$ defined in (3.14). To prove these properties, one can proceed in at least two ways: (i) using the minimization structure of the cell problems (see e.g. Exercise 4.3 in Alouges (2016)), or (ii) proceeding via a direct calculation (see e.g. Proposition 3.2 in Hornung (1997)).

We follow here route (ii). We start with hiding the index j in expression (3.14), allowing us to write

$$\begin{aligned} D_{ik} &:= \sum_{j=1}^{d} \int_Y \left(A_{ik}(y) + A_{ij}(y)\partial_{y_j} W_k(y)\right) dy \\ &= \int_Y A(y) \left(e_k + \nabla_y W_k(y)\right) \cdot e_i dy. \end{aligned}$$

For any $j \in \{1, \ldots, d\}$, we recall that the weak formulation of

$$\nabla_y \cdot (A(y)\nabla_y W_j(y)) = -\nabla_y \cdot (A(y)e_j) \quad (\text{for } y \in Y) \tag{3.15}$$

reads

$$\int_Y A(y) \left(\nabla_y W_j(y) + e_j\right) \nabla_y \phi(y) dy = 0$$

for any test function $\phi \in H^1_\#(Y)$. Take now in (3.15) the test function $\phi := W_i \in H^1_\#(Y)$ to obtain

$$0 = \int_Y A(y)\left(e_j + \nabla_y W_j(y)\right) \nabla_y W_i(y) dy$$

$$= \int_Y A(y) \nabla_y W_j(y) \nabla_y W_i(y) dy + \int_Y A(y) e_j \cdot \nabla_y W_i(y) dy.$$

Hence, relying as well on the symmetry of A, we get

$$\int_Y A(y) e_j \cdot \nabla_y W_i(y) dy = -\int_Y A(y) \nabla_y W_j(y) \nabla_y W_i(y) dy$$

$$= -\int_Y A(y) \nabla_y W_i(y) \nabla_y W_j(y) dy$$

$$= \int_Y A(y) e_i \cdot \nabla_y W_j(y) dy.$$

So, the tensor $\mathbb{D}$ is symmetric.

As we have just seen that $\int_Y A(y)\left(e_j + \nabla_y W_j(y)\right) \nabla_y W_i(y) dy = 0$, we can write

$$D_{ij} = \int_Y A(y)\left(e_j + \nabla_y W_j(y)\right) \cdot e_i dy$$

$$= \int_Y A(y)\left(e_j + \nabla_y W_j(y)\right) \cdot e_i dy + \int_Y A(y)\left(e_j + \nabla_y W_j(y)\right) \nabla_y W_i(y) dy$$

$$= \int_Y A(y)\left(e_j + \nabla_y W_j(y)\right) \cdot \left(e_i + \nabla_y W_i(y)\right) dy.$$

The ellipticity of the tensor $\mathbb{D}$ is now a direct consequence of the ellipticity of A. The reader may wish to give a short direct proof for this last statement.

3.1.3 *Summary of the averaged objects: Upscaled model equation and effective coefficients*

We recall here the result of the averaging process performed by employing the two-scale convergence concept:

The upscaled problem reads: Find $u_0 \in H_0^1(\Omega)$ satisfying the equation

$$\operatorname{div}_x(-\mathbb{D}\nabla_x u_0(x)) = f(x) \quad \text{for } x \in \Omega, \tag{3.16}$$

with the boundary condition

$$u_0(x) = 0 \quad \text{for } x \in \partial\Omega. \tag{3.17}$$

The effective diffusion tensor $\mathbb{D} = [D_{ik}]$ is defined component-wise via

$$D_{ik} := \left[\sum_{j=1}^{d} \int_Y (A_{ik} + A_{ij}\partial_{y_j} W_k(y))\, dy\right] \tag{3.18}$$

for all $(i, k) \in \{1, \ldots, d\} \times \{1, \ldots, d\}$.

We note that this is precisely the same upscaled system as was obtained via the formal homogenization procedure reported in the previous chapter. Interestingly, the two methodologies agree with each other, although they rely on different working assumptions.

3.2 A Corrector Estimate

In some cases, it is possible to estimate theoretically how fast (u_ε) convergences to u_0. Such information can be used to build a computable indicator that can quantify the quality of the averaging.

3.2.1 *How good is the averaging method?*

At this point, we have an averaged (homogenized) model with a computable effective transport coefficient. Inherently, the following unavoidable question arises: *How much information have we lost via averaging?* Looking at the structure of the homogenization *ansatz*

$$u_\varepsilon = u_0 + \varepsilon u_1 + \mathcal{O}(\varepsilon) \tag{3.19}$$

makes us expect something like

$$\frac{u_\varepsilon - u_0}{\varepsilon} = u_1 + \mathcal{O}(\varepsilon).$$

In other words, we expect that showing $u_\varepsilon \to u_0$ (strongly in $L^2(\Omega)$) is related to controlling the growth of u_1. In other words, the golden rule is: the smaller ε is, the better the averaging.

Recall that the two-scale limit problem determines the unique pair $(u_0, u_1) \in H_0^1(\Omega) \times L^2(\Omega; H^1_\#(Y)/\mathbb{R})$. We claim that the limit functions u_0 and u_1 are precisely the same as those arising in (3.19). We have seen earlier that $\nabla u_\epsilon \overset{2}{\rightharpoonup} \nabla_x u_0 + \nabla_y u_1$ in $L^2(\Omega; L^2_\#(Y))$. Can we turn this weak type of convergence into a strong convergence result? In general, this is not possible, but for our precise context, the answer is positive, provided *we extract some oscillations*[8] from the quantity $\nabla u_\varepsilon - \nabla_x u_0$ to get the desired behavior. But how much oscillations should we then extract? Inspired by Lukkassen *et al.* (2002) and Ciorănescu and Donato (1999), we give the precise answer in the forthcoming Theorem 3.1. We refer the reader also to the recent work by Nika and Muntean (2023), where the authors discuss opportunities to relax the regularity restrictions indicated in Theorem 3.1 by suitably regularizing the differential operator.

3.2.2 *A strong convergence result: A first step toward quantitative corrector estimates*

In this section, we prove the following strong convergence result.

Theorem 3.1. *Take $u_\varepsilon \in H_0^1(\Omega)$ as the solution of the target microscopic problem. Let $1 \leq s, t \leq \infty$ such that $\frac{1}{s} + \frac{1}{t} = 1$. Assume that for all $(i, j) \in \{1, \ldots, d\}^2$, we have*

$$\partial_{x_i} u_0 \in L^{2s}(\Omega), \quad \partial_{y_j} W_i \in L^{2t}_\#(Y). \tag{3.20}$$

Then, the following convergence holds:

$$\nabla u_\varepsilon - \nabla_x u_0 - \nabla_y u_1\left(\cdot, \frac{\cdot}{\varepsilon}\right) \to 0 \quad \text{in } L^2(\Omega) \quad \text{as } \varepsilon \to 0. \tag{3.21}$$

---

[8]For a given microscopic problem, it is not quite clear *a priori* how much oscillations can one actually extract. Intuitively, the more oscillations one can control, the better the convergence is expected to be. Readers browsing attentively the literature will observe that the users of periodic unfolding techniques can use their unfolding ideas to derive better results. They do not obtain necessarily faster rates, only the class of microstructures that can be handled is larger. One requires then less regularity (than stated in Theorem 3.1) of the shapes of the microstructures.

Proof. The proof of this statement goes as follows: Using the coercivity condition on the matrix $A\left(\frac{x}{\varepsilon}\right)$, we get

$$\begin{aligned}
&\alpha \int_\Omega |\nabla u_\varepsilon - \nabla_x u_0 - \nabla_y u_1|^2 dx \\
&\quad \leq \int_\Omega A\Big(\frac{x}{\varepsilon}\Big) \left(\nabla u_\varepsilon - \nabla_x u_0 - \nabla_y u_1\right) \left(\nabla u_\varepsilon - \nabla_x u_0 - \nabla_y u_1\right) \\
&\quad = \int_\Omega A\Big(\frac{x}{\varepsilon}\Big) \nabla u_\varepsilon \nabla u_\varepsilon dx - \int_\Omega \nabla_x u_\varepsilon \left(A + A^t\right) \Big(\frac{x}{\varepsilon}\Big) (\nabla_x u_0 + \nabla_y u_1) dx \\
&\qquad + \int_\Omega \Big[A\Big(\frac{x}{\varepsilon}\Big)(\nabla_x u_0 + \nabla_y u_1)\Big] (\nabla_x u_0 + \nabla_y u_1) dx \\
&\quad = \sum_{k=1}^{3} I_k,
\end{aligned}$$

where the terms I_k $(k \in \{1, 2, 3\})$ are defined as

$$\begin{aligned}
I_1 &:= \int_\Omega A\Big(\frac{x}{\varepsilon}\Big) \nabla u_\varepsilon \nabla u_\varepsilon dx, \\
I_2 &:= - \int_\Omega \nabla u_\varepsilon \left(A + A^t\right) \Big(\frac{x}{\varepsilon}\Big) (\nabla_x u_0 + \nabla_y u_1) dx, \\
I_3 &:= \int_\Omega \Big[A\Big(\frac{x}{\varepsilon}\Big)(\nabla_x u_0 + \nabla_y u_1)\Big] (\nabla_x u_0 + \nabla_y u_1) dx.
\end{aligned}$$

Next, we need to pass to the limit $\varepsilon \to 0$ in each of the terms. To do so, we proceed in the following fashion. By the weak formulation of (P_ε), we know that

$$\int_\Omega A\Big(\frac{x}{\varepsilon}\Big) \nabla u_\varepsilon(x) \nabla u_\varepsilon(x) dx = \int_\Omega f(x) u_\varepsilon(x) dx.$$

Passing in this inequality to the two-scale limit $\varepsilon \to 0$, we get that I_1 converges to $\int_\Omega f(x) u_0(x) dx$ as $\varepsilon \to 0$. Now, using the symmetry of A, we can rewrite I_2 as

$$\begin{aligned}
I_2 &= -2 \int_\Omega \nabla u_\varepsilon A\Big(\frac{x}{\varepsilon}\Big) (\nabla_x u_0 + \nabla_y u_1) dx \\
&\overset{\varepsilon \to 0}{\longrightarrow} -2 \int_\Omega \int_Y A(y) [\nabla_x u_0(x) + \nabla_y u_1(x, y)] (\nabla_x u_0(x) + \nabla_y u_1(x, y)) dx dy.
\end{aligned}$$

Similarly, as $\varepsilon \to 0$, we see that I_3 goes to

$$\int_\Omega \int_Y A(y)[\nabla_x u_0(x) + \nabla_y u_1(x,y)](\nabla_x u_0(x) + \nabla_y u_1(x,y))dxdy.$$

Combing the above relations, we get using the weak formulation of the two-scale limit problem (P_0) that

$$\begin{aligned}
0 &\leq \lim_{\varepsilon\to 0} \alpha \int_\Omega \left|\nabla u_\varepsilon - \nabla_x u_0 - \nabla_y u_1\Big(x, \frac{x}{\varepsilon}\Big)\right|^2 dx \\
&\leq \int_\Omega f u_0 \,\mathrm{d}x - \int_\Omega \int_Y A(y)\big[\nabla_x u_0 + \nabla_y u_1(x,y)\big]\big(\nabla_x u_0 + \nabla_y u_1(x,y)\big)dxdy \\
&= 0.
\end{aligned}$$

□

The homogenization community typically calls "corrector" a result of type (3.21). Such a result clarifies how much oscillations need to be subtracted from ∇u_ε to guarantee uniform convergence in $L^2(\Omega)$. However, from a more pragmatic point of view, this information does not tell anything about how fast such strong convergence will be achieved. This type of information is needed to design convergent multiscale numerical methods. More work needs to be done to derive quantitative "corrector estimates", i.e. upper bounds on convergence rates expressed in terms of computable quantities. Once available, such inequalities indicate *how fast* one can approximate u_ε and ∇u_ε in terms of u_0 and $\nabla_x u_0 + \nabla_y u_1(\cdot, \frac{\cdot}{\varepsilon})$, respectively.

We will briefly return to the topic of correctors in Section 4.6.

3.3 Exercises

Exercise 3.3.1. Give a complete proof of the statements from Step 5 of the homogenization procedure mentioned in Section 3.1.1, i.e. ensure the well-posedness of the limit two-scale problem (P_0).

Exercise 3.3.2. Show that $\varphi(x) = A^t\left(\frac{x}{\varepsilon}\right)\nabla_x \phi_1(x, \frac{x}{\varepsilon})$ (for a.e. $x \in \Omega$) is allowed as a test function in the two-scale convergence, as used in Step 4 of the homogenization procedure detailed in Section 2.3.

Exercise 3.3.3. Establish the maximal regularity of the cell functions $W_j(\cdot)$ for all $j \in \{1, \ldots, d\}$.

Exercise 3.3.4. Formulate a variant of Lemma 2.34 (Pavliotis and Stuart, 2008, pp. 26–27) suitable for our setting.

Exercise 3.3.5. Why is the constant $\frac{c_P}{\alpha}$ from (3.5) independent of the choice of ε?

Exercise 3.3.6. Prove (3.6).

Exercise 3.3.7. Let Ω be a bounded non-empty domain in $\mathbb{R}^d$. Consider the following problem, say (P_ε): Find the couple $(u_\varepsilon, v_\varepsilon)$ satisfying the following system of equations:

$$\begin{aligned} \operatorname{div}(-a_\varepsilon(x)\nabla u_\varepsilon(x)) &= k_\varepsilon(x)(-f_\varepsilon(x) + Hv_\varepsilon(x)) \quad \text{for } x \in \Omega, \\ \operatorname{div}(-b_\varepsilon(x)\nabla v_\varepsilon(x)) &= -k_\varepsilon(x)(-f_\varepsilon(x) + Hv_\varepsilon(x)) \quad \text{for } x \in \Omega, \\ u_\varepsilon(x) = \varepsilon &= v_\varepsilon(x) \quad \text{for } x \in \partial\Omega. \end{aligned}$$

Assume $H > 0$, $k_\varepsilon \in L^\infty_+(\Omega)$, $f_\varepsilon \in L^2(\Omega)$, where $f_\varepsilon(x) = f(x, \frac{x}{\varepsilon}) := f_0(x) + \varepsilon^3 f_1(x, \frac{x}{\varepsilon})$ and $k_\varepsilon(x) := k(\frac{x}{\varepsilon})$, as well as $a_\varepsilon(x) := A(\frac{x}{\varepsilon})$ and $b_\varepsilon(x) := B(\frac{x}{\varepsilon})$, with A, B $d \times d$-matrices and $x \in \Omega$. All coefficients are assumed to be Y periodic in the fast variable.

(i) Make meaningful assumptions on the data and parameters of the problem so that the Lax–Milgram lemma ensures weak solutions in $H^1(\Omega) \times H^1(\Omega)$.
(ii) Perform the homogenization asymptotics $\varepsilon \to 0$ using the concept of formal two-scale asymptotics. Derive explicitly the strong formulation of the limit (macroscopic) system.
(iii) Redo the homogenization asymptotics now using the concept of two-scale convergence. Compare the results with what you have obtained with the previous method.
(iv) (Numerical corrector estimates) Consider the case of $\Omega := (0, 1)$. Make a choice of functions/numerical values for k_ε, H, f_0, and f_1. Select $\varepsilon \in \{\frac{1}{10}, \frac{1}{20}, \frac{1}{40}, \frac{1}{100}\}$, and then calculate numerically the following quantities: $\int_\Omega |u_\varepsilon(x) - u_0(x)|^2 dx$, $\int_\Omega |v_\varepsilon(x) - v_0(x)|^2 dx$, $\int_\Omega |\nabla u_\varepsilon(x) - \nabla u_0(x) - \nabla_y u_1(x, \frac{x}{\varepsilon})|^2 dx$, and $\int_\Omega |\nabla v_\varepsilon(x) - \nabla v_0(x) - \nabla_y v_1(x, \frac{x}{\varepsilon})|^2 dx$.
(v) Does problem (P_ε) conserve mass[9] for a fixed choice of ε? What about the limit problem (P_0)?

[9]The question about "conserving the mass" perhaps makes more sense for time-dependent problems. However, one can still imagine this setting as the stationary picture of a time-dependent situation. So, it is about looking at quantities such as $\int_\Omega u_\varepsilon(x)dx$, $\int_\Omega v_\varepsilon(x)dx$, $\int_\Omega u_0(x)dx$, and $\int_\Omega v_0(x)dx$, which are referred here as "mass". Prove that all these quantities are bounded.

Exercise 3.3.8. Let Ω be a bounded non-empty domain in $\mathbb{R}^d$. Consider the following problem, say (P_ε): Find the couple $(u_\varepsilon, v_\varepsilon)$ satisfying the following system of equations:

$$\operatorname{div}(-a_\varepsilon(x)\nabla u_\varepsilon(x)) = k_\varepsilon(x)(-u_\varepsilon(x) + Hv_\varepsilon(x)) \quad \text{for } x \in \Omega,$$

$$\operatorname{div}(-a_\varepsilon(x)\nabla v_\varepsilon(x)) = -k_\varepsilon(x)(-u_\varepsilon(x) + Hv_\varepsilon(x)) \quad \text{for } x \in \Omega,$$

$$u_\varepsilon(x) = \varepsilon = v_\varepsilon(x) \quad \text{for } x \in \partial\Omega.$$

Assume $H > 0$, $k_\varepsilon \in L^\infty_+(\Omega)$, where $k_\varepsilon(x) := k(\frac{x}{\varepsilon})$, as well as $a_\varepsilon(x) := A(\frac{x}{\varepsilon})$, with A a $d \times d$-matrix and $x \in \Omega$. All coefficients are assumed to be Y periodic in the fast variable.

(i) Make meaningful assumptions on the data and parameters of the problem so that the Lax–Milgram lemma can be applied to ensure the weak solvability in $H^1(\Omega) \times H^1(\Omega)$ of (P_ε).

(ii) Perform the homogenization asymptotics $\varepsilon \to 0$ using the concept of formal two-scale asymptotics. Derive explicitly the strong formulation of the limit (macroscopic) system.

(iii) Redo the homogenization asymptotics now using the concept of two-scale convergence. Compare the results with what you have obtained with the formal two-scale asymptotic homogenization method introduced previously.

(iv) (Numerical corrector estimates) Consider the case of $\Omega := (0,1)$. Make a choice of functions/numerical values for k_ε, H, f_0, and f_1. Select $\varepsilon \in \{\frac{1}{10}, \frac{1}{20}, \frac{1}{40}, \frac{1}{100}\}$, and then calculate numerically the following quantities: $\int_\Omega |u_\varepsilon(x) - u_0(x)|^2 dx$, $\int_\Omega |v_\varepsilon(x) - v_0(x)|^2 dx$, $\int_\Omega |\nabla u_\varepsilon(x) - \nabla u_0(x) - \nabla_y u_1(x, \frac{x}{\varepsilon})|^2 dx$, and $\int_\Omega |\nabla v_\varepsilon(x) - \nabla v_0(x) - \nabla_y v_1(x, \frac{x}{\varepsilon})|^2 dx$.

(v) Take $H = 1$. Prove corrector estimates for the homogenization process of the pair $(u_\varepsilon, v_\varepsilon)$.

Exercise 3.3.9. Derive formally the macroscopic equation and corresponding effective coefficients for the following microscopic problem, say (P_ε):

$$\operatorname{div}(-a_\varepsilon(x)\nabla u_\varepsilon) = f(x) \quad \text{for all } x \in \Omega,$$

$$u_\varepsilon(x) = u_D(x/\varepsilon) \quad \text{for all } x \in \partial\Omega,$$

where $\Omega \subset \mathbb{R}^d$, the matrix $a_\epsilon(x) = A(\frac{x}{\varepsilon})$ has periodic entries, the Dirichlet data u_D is sufficiently smooth, and f is not oscillatory (in particular, f is uniform in ϵ). Assume that (P_ϵ) is a well-posed problem, and perform the passage to the formal homogenization limit as $\varepsilon \to 0$. Identify suitable conditions on the initial data $u_D(\cdot)$ to be able to prove rigorously the passage to the limit $\varepsilon \to 0$.

Exercise 3.3.10. Let Ω be a bounded non-empty domain in $\mathbb{R}^d$. Consider the following problem, say (P_ε): Find u_ε satisfying the model equations

$$\operatorname{div}(-a_\varepsilon(x)\nabla u_\varepsilon(x)) + k_\varepsilon(x)u_\varepsilon(x) = f_\varepsilon(x) \quad \text{for } x \in \Omega,$$

$$u_\varepsilon(x) = \varepsilon \quad \text{for } x \in \partial\Omega.$$

Assume $k_\varepsilon \in L^\infty_+(\Omega)$, $f_\varepsilon \in L^2(\Omega)$ (both uniformly bounded), where $f_\varepsilon(x) = f(x, \frac{x}{\varepsilon}) := f_0(x) + \varepsilon f_1(x, \frac{x}{\varepsilon})$ and $k_\varepsilon(x) := k(\frac{x}{\varepsilon})$, as well as $a_\varepsilon(x) := A(\frac{x}{\varepsilon})$ is a $d \times d$-matrix, with uniform bounded entries for each $x \in \Omega$. Y-periodicity is assumed for all functions in the variable x/ε.

(i) Make meaningful assumptions on the data and parameters of the problem so that the Lax–Milgram lemma ensures weak solutions in $H^1(\Omega)$.
(ii) Perform the homogenization asymptotics for $\varepsilon \to 0$ using the concept of two-scale convergence. Derive explicitly the strong formulation of the limit macroscopic system.
(iii) When can problem (P_ε) conserve mass for a fixed choice of ε? What about the limit problem (P_0)? Can it conserve mass?

Exercise 3.3.11. Let Ω be a bounded non-empty domain in $\mathbb{R}^d$. Consider the following problem, say (P_ε): Find u_ε satisfying the model equations

$$\operatorname{div}(-a_\varepsilon(x)\nabla u_\varepsilon(x)) + k_\varepsilon(x)u_\varepsilon(x) = f_\varepsilon(x) \quad \text{for } x \in \Omega,$$

$$u_\varepsilon = 0 \quad \text{at } \Gamma^D,$$

$$-a_\varepsilon(x)\nabla u_\varepsilon \cdot n = g(x) \quad \text{at } \Gamma^N,$$

where $\Gamma^N \cup \Gamma^D = \partial\Omega$ and $\Gamma^N \cap \Gamma^D = \emptyset$. Assume $g \in L^2(\Gamma^N)$. Take $k_\varepsilon \in L^\infty_+(\Omega)$, $f_\varepsilon \in L^2(\Omega)$ (both uniformly bounded), where $f_\varepsilon(x) = f(x, \frac{x}{\varepsilon}) := f_0(x) + \varepsilon f_1(x, \frac{x}{\varepsilon})$ and $k_\varepsilon(x) := k(\frac{x}{\varepsilon})$, as well as $a_\varepsilon(x) := A(\frac{x}{\varepsilon})$ is a $d \times d$-matrix, with uniform bounded entries for each $x \in \Omega$. Y-periodicity is assumed for all functions in the variable x/ε.

(i) Make meaningful assumptions on the data and parameters of the problem so that the Lax–Milgram lemma ensures weak solutions in $H^1(\Omega)$.
(ii) Perform the homogenization asymptotics $\varepsilon \to 0$ using the concept of two-scale convergence. Derive explicitly the strong formulation of the limit macroscopic system.
(iii) Can you show that u_ε lies in $L^1(\Omega)$ uniformly? What about controlling u_0 in the same space?

Exercise 3.3.12. Let Ω be a bounded domain in $\mathbb{R}^d$ ($d \in \{2,3\}$) with the Lipschitz boundary $\partial\Omega$ and $\varepsilon \in (0,+\infty)$ being a parameter. Consider the following linear elliptic equation with oscillating coefficients, which we refer here as (P_ε): Find the unknown u_ε such that it holds

$$\operatorname{div}\big(-a_\varepsilon(x)\nabla u_\varepsilon + \vec{B}_\varepsilon u_\varepsilon\big) = -k_\varepsilon(x)(u_\varepsilon(x) + f_\varepsilon(x)) \quad \text{in } \Omega,$$

$$u_\varepsilon = \varepsilon^d \quad \text{at } \partial\Omega.$$

Assume $f_\varepsilon \in L^2(\Omega)$ uniformly bounded with respect to ε. In this context, the matrix $a_\varepsilon(x) = A(\frac{x}{\varepsilon})$, the vector $\vec{B}_\varepsilon(x) = \vec{B}(\frac{x}{\varepsilon})$, together with the scalars $k_\varepsilon(x) = k(\frac{x}{\varepsilon})$ and $f_\varepsilon(x) = f(\frac{x}{\varepsilon})$ are 1-periodic functions.

(i) Specify the conditions needed for $a_\varepsilon, k_\varepsilon, \vec{B}_\varepsilon$ so that the Lax–Milgram lemma ensures the existence and uniqueness of a weak solution u_ε to (P_ε). Justify your statement.
(ii) Prove that u_ε is uniformly bounded in a suitable Sobolev space (where the passage to the homogenization limit can be performed).
(iii) Pass in (P_ε) via two-scale convergence to the limit $\varepsilon \to 0$ and determine the weak form of the two-scale limit problem, say (P_0) with solution u_0.
(iv) Prove the existence and uniqueness of the weak solution to (P_0).
(v) Get the strong formulation of (P_0). Eliminate u_1.
(vi) Argue in two different ways why $u_0 \in L^1(\Omega)$, i.e. the homogenized system contains "finite mass".
(vii) Extend your list of assumptions from (i) so that you can prove that

$$\nabla u_\varepsilon - \nabla_x u_0 - \nabla_y u_1\Big(\cdot, \frac{\cdot}{\varepsilon}\Big) \to 0 \quad \text{in } L^2(\Omega) \quad \text{as } \varepsilon \to 0.$$

Show the arguments of your proof.

Exercise 3.3.13. Let $S = (0,T)$ be a time interval and $\emptyset \neq \Omega \subset \mathbb{R}^d$ be a space domain. We consider the following parabolic problem modeling a diffusion-chemical reaction process in the homogeneous domain $\Omega_T := S \times \Omega$, given by

$$b_\varepsilon(x)\partial_t u_\varepsilon(t,x) + \operatorname{div}\big(-a_\varepsilon(x)\nabla u_\varepsilon(t,x)\big) = c_\varepsilon(x)f(t,x) \quad \text{for } (t,x) \in \Omega_T$$

$$u_\varepsilon(t,x) = 0 \quad \text{for } (t,x) \in S \times \partial\Omega$$

$$u_\varepsilon(0,x) = d(x) \quad \text{for } x \in \bar{\Omega},$$

where $a_\varepsilon(x) = A(\frac{x}{\varepsilon})$ satisfies the usual coercivity assumption, $b(\cdot)$ is a bounded positive Y-periodic function such that

$$0 < b^- \leq b(y) \leq b^+ < +\infty$$

for all $y \in Y$, $c \in L^2_\#(Y)$, $d \in L^2(\Omega)$, and $f \in L^2(\Omega_T)$. Use the concept of two-scale convergence adapted to time-dependent settings to derive the homogenized version of this problem.

3.4 Solutions

Solution 3.3.1. As a result of the passage to the two-scale convergence limit, we have obtained that the pair

$$(u_0, u_1) \in H^1_0(\Omega) \times L^2(\Omega; H^1_\#(Y)/\mathbb{R})$$

satisfies the identity

$$\frac{1}{|Y|}\int_{\Omega\times Y} A(y)(\nabla_x u_0 + \nabla_y u_1)(\nabla_x \phi_0 + \nabla_y \phi_1)dxdy = \frac{1}{|Y|}\int_{\Omega\times Y} f\phi_0 dxdy$$

for all test functions $(\phi_0, \phi_1) \in C^\infty_0(\Omega) \times C^\infty_0(\Omega; C^\infty_\#(Y))$.

If we now perceive the set $C^\infty_0(\Omega) \times C^\infty_0(\Omega; C^\infty_\#(Y))$ as being dense in $H^1_0(\Omega) \times L^2(\Omega; H^1_\#(Y)/\mathbb{R})$, then we can pose the weak formulation of the target problem as follows: Find the pair

$$(u_0, u_1) \in H^1_0(\Omega) \times L^2(\Omega; H^1_\#(Y)/\mathbb{R})$$

satisfying the identity

$$\int_{\Omega\times Y} A(y)(\nabla_x u_0 + \nabla_y u_1)(\nabla_x \phi_0 + \nabla_y \phi_1)dxdy = \int_\Omega f\phi_0 dx$$

for all test functions $(\phi_0, \phi_1) \in H^1_0(\Omega) \times L^2(\Omega; H^1_\#(Y)/\mathbb{R})$. To obtain the last identity, we used the fact that $|Y| = 1$ and that the functions f and u_0 are in fact independent on $y \in Y$.

As the next step, we verify extent to which the hypotheses of the Lax–Milgram lemma (see Theorem 7.1) fit our scenario. The complete details of the solvability proof can be found in the work of, e.g. Alouges (2016) and Lukkassen *et al.* (2002). We provide these details here again as we think this is useful information – the structures of the involved functionals are typical for the homogenization of linear elliptic problems. Similar structures will appear again in those concrete applications of the methodology where elliptic equations are part of the model to be homogenized.

It is convenient to introduce the Hilbert space

$$\mathcal{H} := H^1_0(\Omega) \times L^2(\Omega; H^1_\#(Y)/\mathbb{R})$$

endowed with the norm

$$||(u_0, u_1)||_{\mathcal{H}} := ||\nabla_x u_0||_{L^2(\Omega)} + ||\nabla_y u_1||_{L^2(\Omega\times Y)}.$$

We define the mappings $\mathcal{A} : \mathcal{H} \times \mathcal{H} \to \mathbb{R}$ and $\mathcal{F} : \mathcal{H} \to \mathbb{R}$ by means of the expressions

$$\begin{aligned}
\mathcal{A}[(u_0, u_1), (\phi_0, \phi_1)] &:= \int_{\Omega\times Y} A(y)\nabla_x u_0(x)(\nabla_x \phi_0(x) + \nabla_y \phi_1(x, y))dxdy \\
&\quad + \int_{\Omega\times Y} A(y)\nabla_y u_1(x, y)(\nabla_x \phi_0(x) + \nabla_y \phi_1(x, y))dxdy, \\
\mathcal{F}[(\phi_0, \phi_1)] &:= \int_\Omega f(x)\phi_0(x)dx
\end{aligned}$$

for all $(\phi_0, \phi_1) \in \mathcal{H}$.

Now, we can reformulate our problem as follows: Find the pair $(u_0, u_1) \in \mathcal{H}$ such that

$$\mathcal{A}[(u_0, u_1), (\phi_0, \phi_1)] = \mathcal{F}[(\phi_0, \phi_1)] \tag{3.22}$$

for all $(u_0, u_1) \in \mathcal{H}$.

It is straightforward to see that $\mathcal{A}$ is symmetric (if $A(\cdot)$ is symmetric) and bilinear and also that $\mathcal{F}$ is linear. The continuity property of both $\mathcal{A}$ and $\mathcal{F}$ is linked to their boundedness. We observe that

$$\begin{aligned}
&|\mathcal{A}[(u_0, u_1), (\phi_0, \phi_1)]| \\
&\quad = \left| \int_{\Omega\times Y} A(y)\nabla_x u_0(x) + \nabla_y u_1(x, y))(\nabla_x \phi_0(x) + \nabla_y \phi_1(x, y))dxdy \right. \\
&\quad \leq \int_{\Omega\times Y} |A(y)||\nabla_x u_0(x) + \nabla_y u_1(x, y)||(\nabla_x \phi_0(x) + \nabla_y \phi_1(x, y))dxdy \\
&\quad \leq \|A\|_{L^\infty(Y)}\|(u_0, u_1)\|_{\mathcal{H}}\|(\phi_0, \phi_1)\|_{\mathcal{H}}
\end{aligned}$$

and

$$\begin{aligned}
|\mathcal{F}[((\phi_0, \phi_1)]| &= \left| \int_\Omega f(x)\phi_0(x)dx \right| \\
&\leq \int_\Omega |f(x)||\phi_0(x)|dx \leq \|f\|_{L^2(\Omega)}\|\phi_0\|_{L^2(\Omega)} \\
&\leq c_P\|f\|_{L^2(\Omega)} \left(\|\nabla_x \phi_0\|_{L^2(\Omega)} + \|\nabla_y \phi_1\|_{L^2(\Omega\times Y)}\right) \\
&\leq c_P\|f\|_{L^2(\Omega)}\|(\phi_0, \phi_1)\|_{\mathcal{H}}.
\end{aligned}$$

It remains to prove the coercivity of $\mathcal{A}(\cdot,\cdot)$, i.e. there exists $c > 0$ such that

$$\mathcal{A}[(u_0,u_1),(u_0,u_1)] \geq c\|(u_0,u_1)\|_{\mathcal{H}}^2.$$

By direct calculations, we can see that the ellipticity of the matrix $A(\cdot)$ leads to the ellipticity of the bilinear form $\mathcal{A}(\cdot,\cdot)$. To observe this aspect, we proceed in the following manner:

$$\begin{aligned}
&\mathcal{A}[(u_0,u_1),(u_0,u_1)]\\
&\quad = \int_{\Omega\times Y} A(y)[\nabla_x u_0(x) + \nabla_y u_1(x,y)]^2 dxdy\\
&\quad \geq \alpha\left(\int_{\Omega\times Y} |\nabla_x u_0|^2 + \int_{\Omega\times Y} |\nabla_y u_1|^2 + 2\int_{\Omega\times Y} \nabla_x u_0 \nabla_y u_1(x,y)dxdy\right)\\
&\quad = \alpha\left(\int_{\Omega\times Y} |\nabla_x u_0|^2 + \int_{\Omega\times Y} |\nabla_y u_1|^2\right) = \alpha\|(u_0,u_1)\|_{\mathcal{H}}^2.
\end{aligned}$$

To obtain this result, we use Fubini's theorem to see that

$$2\alpha\int_{\Omega\times Y} \nabla_x u_0 \nabla_y u_1(x,y)dxdy = \int_\Omega \nabla_x u_0\left(\int_Y \nabla_y u_1(x,y)dy\right)dx$$

vanishes due to the fact that u_1 is actually a Y-periodic function.

To complete the proof of the well-posedness for the target problem, the stability of the wanted solution with respect to parameters has to be investigated as well. Let $\mathcal{U}^1 := (u_0^1, u_1^1)$ and $\mathcal{U}^2 := (u_0^2, u_1^2)$ be two pairs of weak solutions to the limit two-scale problem (P_0), which correspond to the data (A^1, f^1) and (A^2, f^2), respectively. The stability property is guaranteed if, for some $c > 0$, an estimate of the type

$$\|\mathcal{U}^2 - \mathcal{U}^1\|_{\mathcal{H}} \leq c\left(\|A^2 - A^1\|_{L^\infty(Y)} + \|f^2 - f^1\|_{L^2(\Omega)}\right) \tag{3.23}$$

is available. The stability estimate (3.23) tells us that small changes in the data will lead to small changes in the solution. In particular, it also indicates that one expects a macroscopic output that is stable with respect to the microscopic input. This feature is guaranteed, as we can point out the existence of a constant $c > 0$ such that

$$\|\mathcal{U}^2 - \mathcal{U}^1\|_{\mathcal{H}} \leq c\|A^2 - A^1\|_{L^\infty(Y)}. \tag{3.24}$$

Studying macroscopic effects produced by varying the shape of the microstructure Y is also possible, but this is not the current focus.

We give now a short proof of (3.23). For any pair of test functions $(\phi_0, \phi_1) \in \mathcal{H}$, we write the identities

$$\int_{\Omega\times Y} A^1(y)(\nabla_x u_0^1 + \nabla_y u_1^1)(\nabla_x \phi_0 + \nabla_y \phi_1)dxdy = \int_\Omega f^1 \phi_0 dx,$$

$$\int_{\Omega\times Y} A^2(y)(\nabla_x u_0^2 + \nabla_y u_1^2)(\nabla_x \phi_0 + \nabla_y \phi_1)dxdy = \int_\Omega f^2 \phi_0 dx.$$

Subtracting the first identity from the second one and then choosing afterward the test function

$$(\phi_0, \phi_1) := (u_0^2 - u_0^1, u_1^2 - u_1^1)$$

leads to the following expression:

$$\int_{\Omega\times Y} A^2(y) \left[\nabla_x u_0^2 + \nabla_y u_1^2 - (\nabla_x u_0^1 + \nabla_y u_1^1)\right]^2 dxdy$$
$$= \int_{\Omega\times Y} \left(A^1(y) - A^2(y)\right) \left(\nabla_x u_0^1 + \nabla_y u_1^1\right)$$
$$\times \left[\nabla_x u_0^2 + \nabla_y u_1^2 - (\nabla_x u_0^1 + \nabla_y u_1^1)\right] + \int_\Omega (f^2 - f^1) u_0.$$

Using the ellipticity condition on $A^2(\cdot)$ with constant $\alpha_2 > 0$, we obtain

$$\alpha_2 \int_{\Omega\times Y} \left[\nabla_x u_0^2 + \nabla_y u_1^2 - (\nabla_x u_0^1 + \nabla_y u_1^1)\right]^2 dxdy$$
$$\leq \int_{\Omega\times Y} |A^1(y) - A^2(y)||\nabla_x u_0^1 + \nabla_y u_1^1||\nabla_x u_0^2 + \nabla_y u_1^2 - (\nabla_x u_0^1 + \nabla_y u_1^1)$$
$$+ ||f^2 - f^1||_{L^2(\Omega)}||u_0||$$
$$\leq ||A^1 - A^2||_{L^\infty(Y)}||\nabla_x u_0^1 + \nabla_y u_1^1||_{L^2(\Omega\times Y)}$$
$$\cdot \, ||\nabla_x u_0^2 + \nabla_y u_1^2 - (\nabla_x u_0^1 + \nabla_y u_1^1) \, ||_{L^2(\Omega\times Y)}$$
$$+ \, c_P||f^2 - f^1||_{L^2(\Omega)}||\mathcal{U}^2 - \mathcal{U}^1||_{\mathcal{H}},$$

which implies

$$\alpha_2||\mathcal{U}^2 - \mathcal{U}^1||^2_{\mathcal{H}}$$
$$\leq ||A^1 - A^2||_{L^\infty(Y)}||\mathcal{U}^1||_{\mathcal{H}}||\mathcal{U}^2 - \mathcal{U}^1||_{\mathcal{H}} + c_P||f^2 - f^1||_{L^2(\Omega)}||\mathcal{U}^2 - \mathcal{U}^1||_{\mathcal{H}}.$$

Dividing the obtained result by $||\mathcal{U}^2 - \mathcal{U}^1||_{\mathcal{H}}$ gives

$$||\mathcal{U}^2 - \mathcal{U}^1||_{\mathcal{H}} \leq \left(c_P + ||\mathcal{U}^1||_{\mathcal{H}}\right)\left(||A^1 - A^2||_{L^\infty(Y)} + ||f^2 - f^1||_{L^2(\Omega)}\right).$$

Hence, the constant c mentioned in (3.23) can be taken to be

$$c := c_P + \|\mathcal{U}^1\|_{\mathcal{H}}.$$

Solution 3.3.7. This is a "typical" set of exam questions.

Solution 3.3.8. The reader may first want to use for all $x \in \bar{\Omega}$ the transformation $U_\varepsilon(x) := u_\varepsilon(x) - \varepsilon$ and $V_\varepsilon(x) := v_\varepsilon(x) - \varepsilon$. As an alternative to applying the Lax–Milgram lemma to the posed system, one can reformulate the problem relying on an additional transformation of the type $w_\varepsilon := U_\varepsilon - HV_\varepsilon$ and $p_\varepsilon := U_\varepsilon + V_\varepsilon$. This change of functions leads to two decoupled linear elliptic partial differential equations, which simplifies the structure of the original problem.

Solution 3.3.9. Make the transformation $v := u - u_D$, and work with the partial differential equation reformulated in terms of v.

Solution 3.3.13. The analysis done in this chapter extends in a straightforward way to linear evolution problems. Essentially, one makes use now of Definition 7.7. Note that if $c_\varepsilon(x) = 1$ for all $x \in \Omega$, then this example is treated by Allaire (2012). An alternative approach is presented in Chapter 11 in Ciorănescu and Donato (1999), where Tartar's method of oscillating test functions is used to perform the homogenization asymptotics.

Chapter 4

Handling Porous Media as Periodically Perforated Media: The Homogenization Route

In this chapter, we use the concept of two-scale convergence to perform the periodic homogenization asymptotics for boundary-value problems posed in perforated domains. Such perforated domains are assumed in this context to be good approximations of given regular porous media. The choice of perforations we have in mind are meant to be suitable approximations of typical microstructures arising in materials. In most cases, they can be perceived as pores inside a regular porous media. The underlying porous medium is referred to as "regular" mainly because it is assumed to be made of a distribution of periodically placed pores[1] with sufficiently smooth boundaries. The mathematical challenge lies in capturing, in the homogenization limit, macroscopic effects that can be linked to the choices of pore shapes and microscopic boundary processes, which correspond to production terms that are defined across microscopic interfaces.

4.1 Setting the Stage

We restrict our focus on the usage of the concepts of two-scale convergence for domains and periodically oscillating surfaces to perform the rigorous homogenization of a linear second-order elliptic equation posed in a perforated domain Ω_ε. The presentation of the techniques is done in a similar style as in the previous chapter. Most of the notation is kept unchanged.

[1]The word "pore" is used here with a generic meaning. Depending on the situation at hand, it may need to be replaced by "void" or by "microstructure".

Doing so, we hope that the reader rapidly notices the main differences that appear when homogenizing equations in homogeneous *versus* perforated domains.

Our guiding question is the following:

> How does the presence of perforations, their precise shapes, and eventual boundary productions affect the outcome of the homogenization procedure?

We wish to see, preferably in a quantitative way, how the volume and shape of the microstructure (perforation) enter the outcome of the averaging procedure. Furthermore, we use this occasion to discuss issues connected to the error analysis of the homogenization process (we will look at *correctors*), as well as further suitable scalings. To fix ideas, we consider several scenarios occurring in the modeling of reactive flows through porous media.

4.2 Geometry of Perforations: Definition of the Perforated Set

We define first the basic geometry of the perforations. We do this in the same spirit as in Section 2.3.

To describe the geometry of the perforated Ω_ε, we rely on the following notation [2]: Let $Z \subset \mathbb{R}^d$ be a hypercube of volume one with $Z := [0,1]^d$, $Y := Z \backslash \bar{Y}_0$, where $Y_0 \subset Z$ is an open set with smooth boundary $\Gamma := \partial Y_0$ not intersecting ∂Z. Y_0 is here the hole or void space of the pore. Furthermore, let $k := (k_1, \ldots, k_d) \in \mathbb{Z}^d$ be a vector of indices, and let $e := (e_1, \ldots, e_d)$ be the unit vector in $\mathbb{R}^d$. For a set $X \subset Z$, we recall the notation X^k, which is the shifted subset

$$X^k := X + \sum_{i=1}^{d} k_i e_i.$$

The pore matrix (or pore skeleton) can now be defined by

$$\Omega_0^\varepsilon := \bigcup_{k \in \mathbb{Z}^d} \{\varepsilon Y_0^k : \varepsilon Y_0^k \subset \Omega\},$$

[2]This notation is inspired very much by the one used by Hornung and Jäger (1991).

while the total pore space is

$$\Omega^{\varepsilon} := \Omega \backslash \Omega_0^{\varepsilon}.$$

The total inner surface of the pore matrix is denoted by

$$\Gamma^{\varepsilon} := \partial \Omega_0^{\varepsilon} = \bigcup_{k \in \mathbb{Z}^d} \{\varepsilon \Gamma^k : \varepsilon \Gamma^k \subset \Omega\}.$$

Correspondingly, we introduce the unit normal vector n_{ε} acting on the smooth (oscillating) surface Γ^{ε}. Let $d\sigma_{\varepsilon}$ be the corresponding (oscillating) measure defined on Γ_{ε} and $d\sigma_y$ be the measure defined on Γ.

There are, of course, a few alternative ways to describe subsets of perforated domains. For instance, denoting by $\chi : Z \to \mathbb{R}$ the standard characteristic function, we can benefit from using $\chi(\cdot)$ to indicate the set Y_0 in Z. In this way, we can represent the perforated domain also via

$$\Omega_{\varepsilon} := \left\{ x \in \Omega : \chi\left(\frac{x}{\varepsilon}\right) = 0 \right\} = \left\{ x \in \Omega : \frac{x}{\varepsilon} \in Y \right\}.$$

Consequently, the corresponding microscopic oscillating surface can be seen as

$$\Gamma_{\varepsilon} := \left\{ x \in \Omega : \frac{x}{\varepsilon} \in \Gamma \right\}.$$

It is worth noting that the cell Z is usually identified as the unit torus $\mathbb{R}^d/\mathbb{Z}^d$.

At this stage, we are wondering:

(i) What does it mean that the perforated domain Ω_{ε} converges to the homogeneous (non-perforated) domain $\Omega_0 = \Omega$ as the diameter ε of the holes vanishes?

(ii) What happens with the perforations (microstructures) as $\varepsilon = 0$? Is the microstructure information completely lost during the passage to the limit?

(iii) Can we apply for settings involving perforated domains the same periodic homogenization strategy as employed for the fixed domain case of Ω (with coefficients defined on a period Y). In other words, will we be able to rely on the concept of two-scale convergence when passing to the limit $\varepsilon \to 0$ in the partial differential equations posed on the perforated domain?

We address these questions for a couple of simple scenarios involving linear second-order elliptic equations posed in periodically perforated domains with different conditions imposed across the microscopic boundaries.

4.3 Homogenization of a Linear Second-Order Elliptic Equation Posed in a Periodically Perforated Domain

We consider our model problem posed now in a perforated domain Ω_ε defined as in Section 4.2. What comes next is the simplest situation we can imagine at this stage: a toy setting. This example is inspired from Ciorănescu and Saint Jean Paulin (1979) and Allaire and Murat (1993).

Consider an arbitrarily fixed $\varepsilon > 0$. The microscopic problem reads as follows:

Find $u_\varepsilon \in H^1(\Omega_\varepsilon; \partial\Omega)$ such that

$$\operatorname{div}\big(-a_\varepsilon(x)\nabla u_\varepsilon\big) = f_\varepsilon(x) \quad \text{for } x \in \Omega_\varepsilon, \tag{4.1}$$

$$-a_\varepsilon(x)\nabla u_\varepsilon \cdot n_\varepsilon = 0 \quad \text{at } \Gamma_\varepsilon, \tag{4.2}$$

$$u_\varepsilon = 0 \quad \text{at } \partial\Omega, \tag{4.3}$$

where $a_\varepsilon(x) = A\left(\frac{x}{\varepsilon}\right) \in \mathbb{R}^{d^2}$.

We refer to (4.1)–(4.3) as problem (P_ε).

As one can see from (4.2), this first scenario does not include any boundary productions; therefore, we start with avoiding deliberately all mathematical difficulties related to the treatment of singular productions defined on oscillating boundaries. Consequently, the physics which can be captured by (P_ε) is rather trivial. This is the reason why we call (P_ε) a "toy". However, mathematically, we face here a new interesting problem: The original domain Ω is now perforated and becomes Ω_ε. So, it changes the shape when changing the value of ε. This complicates the asymptotic thinking.

It is worth noting that the way (P_ε) is formulated suggests that the microstructures (perforations) do not touch $\partial\Omega$; compare the left panel in Figure 4.1. However, trusting Allaire and Murat (1993), for this particular

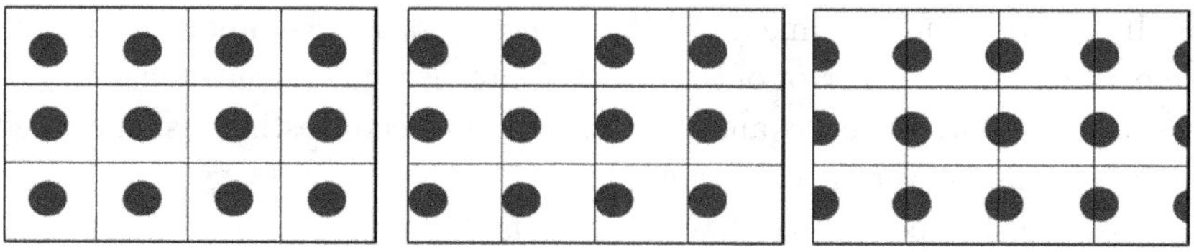

Figure 4.1. Examples of sets Ω_ε with regular and nice perforations (left), perforations touching the boundary of the periodic cell (middle), and with perforations intersecting the boundary of the periodic cell (right).

case, it is actually not disturbing if such touching takes place (like in the middle and right panels in Figure 4.1) provided one requires instead of (4.2)–(4.3), the following boundary conditions:

$$-a_\varepsilon(x)\nabla u_\varepsilon \cdot n_\varepsilon = 0 \quad \text{at } \Gamma_\varepsilon \backslash \partial\Omega, \tag{4.4}$$

$$u_\varepsilon = 0 \quad \text{at } \partial\Omega \cap \Gamma_\varepsilon. \tag{4.5}$$

As in Chapter 3, we consider the following assumptions (H1)–(H4) to be fulfilled. They are slightly adapted to the current setting:

(H1) The matrix A satisfies the coercivity condition

$$\exists \alpha, \beta > 0 \text{ such that } \alpha|\zeta|^2 \leq \sum_{i,j=1}^{d} A_{ij}(y)\zeta_i\zeta_j \leq \beta|\zeta|^2, \tag{4.6}$$

for a.e. $y \in Y$ and for all $\zeta \in \mathbb{R}^d$. The ellipticity constants α and β are taken to be independent of ε.

(H2) $A_{ij} \in L^\infty_\#(Y)$ for all $(i,j) \in \{1,\ldots,d\}^2$, and A is symmetric.

(H3) $f_\varepsilon \in L^2(\Omega_\varepsilon)$ such that there exists $c > 0$ independent of ε such that

$$\|f_\varepsilon\|_{L^2(\Omega_\varepsilon)} \leq c. \tag{4.7}$$

To keep things simple, we imagine that the function $f_\varepsilon(\cdot)$ admits the structure $f_\varepsilon(x) = f_0(x, \frac{x}{\varepsilon})$, where $f_0(\cdot,\cdot)$ is a given function which is sufficiently smooth in the second variable.

(H4) Ω_ε is the periodically perforated domain, where Ω is a parallelepiped in $\mathbb{R}^d$ with suitable rational corner coordinates so that it can be covered perfectly with suitably scaled Y-hypercubes.

If ones does not assume (H4), the homogenization asymptotics can still be performed, but an error of the order of $\mathcal{O}(\varepsilon)$ is typically introduced. This effect can, in principle, be analyzed through corrector estimates, enhanced with boundary-layer-type information.

Up until the point where we pass to the homogenization limit, we can follow the steps of the homogenization procedure outlined in Chapter 3. We split the application of the homogenization procedure into seven steps as follows:

Step 1: Well-posedness of the microscopic problem. As always, we start off with exploring the solvability of the posed problem. Set $V_\varepsilon := H^1(\Omega_\varepsilon; \partial_\Omega)$.

Definition 4.1 (Weak solution to (P_ε)). $u_\varepsilon \in V_\varepsilon$ is a weak solution to (P_ε) if and only if for all $\varphi \in V_\varepsilon$ the following identity holds:

$$(a_\varepsilon \nabla u_\varepsilon, \nabla\varphi) = (f_\varepsilon, \varphi). \tag{4.8}$$

The Lax–Milgram lemma (Theorem 7.1) ensures the existence and uniqueness of $u_\varepsilon \in V_\varepsilon$ solving (P_ε) in the sense of Definition 4.1.

Step 2: Derivation of ε-independent *a priori* estimates. Choosing in (4.8) as the test function $\varphi = u_\varepsilon$ and using the coercivity condition (H1), we obtain

$$\begin{aligned}
\alpha\|u_\varepsilon\|^2_{V_\varepsilon} &\le (a_\varepsilon \nabla u_\varepsilon, \nabla u_\varepsilon) = (f_\varepsilon, u_\varepsilon) \le \|f_\varepsilon\|_{L^2(\Omega_\varepsilon)}\|u_\varepsilon\|_{L^2(\Omega_\varepsilon)} \\
&\le c_P\|f_\varepsilon\|_{L^2(\Omega_\varepsilon)}\|u_\varepsilon\|_{V_\varepsilon} \\
&\le \eta\|u_\varepsilon\|^2_{V_\varepsilon} + c_\eta c_P^2\|f_\varepsilon\|^2_{L^2(\Omega_\varepsilon)},
\end{aligned} \tag{4.9}$$

where the constant $c_P > 0$ is from Poincaré's inequality (note that for any fixed $\varepsilon \ge 0$, we have $V_\varepsilon \hookrightarrow L^2(\Omega_\varepsilon)$). Here, take $\eta > 0$ so that it yields $0 < c_\eta < \infty$. Choosing $\eta \in (0, \alpha)$ in inequality (4.9), we are led to the estimate

$$\|u_\varepsilon\|_{V_\varepsilon} \le c_P\sqrt{\frac{c_\eta}{\alpha - \eta}}\|f_\varepsilon\|_{L^2(\Omega_\varepsilon)}. \tag{4.10}$$

Alternatively, (4.9) can be rewritten as:

$$\begin{aligned}
\alpha\|u_\varepsilon\|^2_{V_\varepsilon} &\le (a_\varepsilon \nabla u_\varepsilon, \nabla u_\varepsilon) = (f_\varepsilon, u_\varepsilon) \le \|f_\varepsilon\|_{L^2(\Omega_\varepsilon)}\|u_\varepsilon\|_{L^2(\Omega_\varepsilon)} \\
&\le \|f_\varepsilon\|_{L^2(\Omega_\varepsilon)}\left(\|u_\varepsilon\|_{L^2(\Omega_\varepsilon)} + \|\nabla u_\varepsilon\|_{L^2(\Omega_\varepsilon)}\right) \\
&\le \|f_\varepsilon\|_{L^2(\Omega_\varepsilon)}\|u_\varepsilon\|_{V_\varepsilon}.
\end{aligned}$$

Relying on the equivalence $\|\cdot\|_{H^1(\Omega_\varepsilon)} \sim \|\cdot\|_{V_\varepsilon}$, there is a constant $\tilde{c} > 0$ such that $\|m\|_{H^1(\Omega_\varepsilon)} \leq \tilde{c}\|m\|_{V_\varepsilon}$ for $m \in V_\varepsilon$. This gives

$$\|u_\varepsilon\|_{V_\varepsilon} \leq \frac{\tilde{c}}{\alpha}\|f_\varepsilon\|_{L^2(\Omega_\varepsilon)}. \tag{4.11}$$

At this point, we may wonder whether the constants c_P and/or $\tilde{c}$ depend on ε. This is generally a significant concern, as the answer is not always affirmative, which can pose serious challenges for the homogenization-limiting procedure. However, one can prove that c_P and $\tilde{c}$ are independent of the choice of ε for selected classes of suitably nice microstructures. For the microstructure Y_0 chosen here and for the proposed boundary conditions, the constant c_P (and hence, also $\tilde{c}$) does not depend on ε; see, e.g. Lemma A4 in Allaire and Murat (1993), clarifying the situation about c_P, which can then be used to elucidate the independence of $\tilde{c}$ from the effects of varying ε.

What to do next? The case of perforated media brings in a couple of new aspects that need to be considered carefully. One of them is the following.

Can we now find subsequences u_ε such that in $L^2(\Omega_\varepsilon)$, and respectively in $H^1(\Omega_\varepsilon)$), the following convergences hold true:

$$u_\epsilon \overset{2}{\rightharpoonup} u_0, \tag{4.12}$$

$$\nabla u_\epsilon \overset{2}{\rightharpoonup} \nabla_x u_0 + \nabla_y u_1 \quad \text{as } \varepsilon \to 0? \tag{4.13}$$

This question is generally difficult to handle. The main reason is that the function spaces $L^2(\Omega_\varepsilon)$, and respectively $H^1(\Omega_\varepsilon)$, vary with changing ε (the underlying Lebesgue measure is variable in terms of ε). To address the matter in a satisfactory way, one additional technical ingredient needs to be available. We address this issue in the next step.

Step 3: Extension to fixed domains. It is rather uncomfortable to study the convergence of the sequence (u_ε) with $\varepsilon \to 0$ when both the sequence of functions and the underlying function space are dependent on the varying parameter ε. Hence, it is convenient to fix the underlying function space. For this to happen, the new ingredient now needed is an extension procedure in Sobolev spaces. In fact, one has to perform the extension of the sequence of oscillating functions (u_ε) defined originally on the space

$H^1(\Omega_\varepsilon : \partial\Omega)$ to some sequence of functions $(\tilde{u}_\varepsilon)$ defined on the "fixed" space $H^1(\Omega)$. Regarding the setting described in (4.1), we can always construct the function $\tilde{u}_\varepsilon$ by extending with zero the function $u_\varepsilon \in H^1(\Omega_\varepsilon; \partial\Omega)$, that is, we take

$$\tilde{u}_\varepsilon(x) := \begin{cases} u_\varepsilon(x) & \text{if } x \in \Omega_\varepsilon, \\ 0 & \text{if } x \in \Omega \backslash \Omega_\varepsilon, \end{cases}$$

or we can write this informally as

$$\tilde{u}_\varepsilon(x) := \chi\left(\frac{x}{\varepsilon}\right) u_\varepsilon(x),$$

where $\chi\left(\frac{x}{\varepsilon}\right)$ is the characteristic function associated with the set Y_0; hence, we can describe the perforated domain as

$$\Omega_\varepsilon = \left\{ x \in \Omega : \chi\left(\frac{x}{\varepsilon}\right) = 0 \right\}.$$

It is worth noting that for this particular case, the zero extension holds true also when the hole Y_0 touches ∂Z, as drawn in Figure 4.1. If the holes do not touch neither the exterior boundary nor each other, then the extension idea proposed by Ciorănescu and Saint Jean Paulin (1979) works here very well.

Besides extending the concept of solution to a fixed domain, the data of the problem needs to be extended accordingly. By (H3), together with (4.7), we deduce that the function $\tilde{f}_\varepsilon : \Omega \to \mathbb{R}$ defined by means of

$$\tilde{f}_\varepsilon(x) = \chi\left(\frac{x}{\varepsilon}\right) f_\varepsilon(x)$$

builds, in fact, a uniformly bounded sequence $\tilde{f}_\varepsilon \in L^2(\Omega)$. We conclude that as $\varepsilon \to 0$, we have

$$\tilde{f}_\varepsilon(x) = \chi\left(\frac{x}{\varepsilon}\right) f_\varepsilon(x) \overset{2}{\rightharpoonup} \chi(y) f_0(x, y).$$

Regarding the extension of $a_\varepsilon \in L^\infty(\Omega_\varepsilon)$ to a fixed domain, we proceed similarly and take its zero extension $\tilde{a}_\varepsilon$ to $L^\infty(\mathbb{R}^d)$.

Recalling (4.11) (or (4.10)), Theorem 7.7, and the convergence properties of $\chi\left(\frac{x}{\varepsilon}\right)$ as $\varepsilon \to 0$, we state that there exists a pair of functions

$$(u_0, u_1) \in H_0^1(\Omega) \times L^2(\Omega; H^1_\#(Y)/\mathbb{R})$$

such that (along subsequences) the following convergences hold:

$$\tilde{u}_\varepsilon(x) = \chi\left(\frac{x}{\varepsilon}\right) u_\varepsilon(x) \overset{2}{\rightharpoonup} \chi(y) u_0(x, y)$$

and

$$\tilde{\nabla} u_\varepsilon(x) = \chi\left(\frac{x}{\varepsilon}\right) \nabla u_\varepsilon(x) \stackrel{2}{\rightharpoonup} \chi(y) \left(\nabla_x u_0(x) + \nabla_y u_1(x,y)\right).$$

See a similar statement in Lemma 4.3 from Ainouz (2007).

Once the extension to function spaces defined on Ω is done, the standard homogenization procedure applies, as for partial differential equations with oscillating coefficients posed on a homogeneous domain, as described in the previous chapter. Note that, in general, such a zero extension does not lie in the extended (fixed) Sobolev space. If that is the case, then other extensions need to be constructed. Using now the information from Step 2, we have just shown that the extension $\tilde{u}_\varepsilon$ is uniformly bounded in $H_0^1(\Omega)$. Consequently, the homogenization procedure takes the familiar route through Steps 4 and 5: Step 4 – compactness step, Step 5 – weak formulation of the limit two-scale problem, Step 6 – weak solvability of (P_0)), and finally, Step 7 – strong formulation of the two-scale limit problem).

Step 4 and Step 5: Passage to $\varepsilon \to 0$ and derivation of the limit two-scale problem. We can now rewrite (4.8) as

$$\int_\Omega a_\varepsilon(x)\chi\left(\frac{x}{\varepsilon}\right) \nabla u_\varepsilon(x) \nabla \psi(x) dx = \int_\Omega \chi\left(\frac{x}{\varepsilon}\right) f_\varepsilon(x)\psi(x) dx \tag{4.14}$$

for all $\psi \in H_0^1(\Omega)$. Now, we proceed as in the previous chapter. We set $x \in \Omega$ arbitrarily and choose in (4.14) as test function

$$\psi(x) = \phi_0(x) + \varepsilon\phi_1\left(x, \frac{x}{\varepsilon}\right),$$

with

$$(\phi_0, \phi_1) \in C_0^\infty(\Omega) \times C_0^\infty(\Omega; C_\#^\infty(Y)).$$

We obtain

$$\begin{aligned}\int_\Omega A\Big(\frac{x}{\varepsilon}\Big)\chi\left(\frac{x}{\varepsilon}\right) \nabla u_\varepsilon(x)\bigg(\nabla\phi_0(x) + \nabla_y\phi_1\Big(x, \frac{x}{\varepsilon}\Big) + \varepsilon\nabla_x\phi_1\Big(x, \frac{x}{\varepsilon}\Big)\bigg)dx \\ = \int_\Omega \chi\left(\frac{x}{\varepsilon}\right) f_\varepsilon(x)\phi_0(x)dx + \varepsilon\int_\Omega \chi\left(\frac{x}{\varepsilon}\right) f_\varepsilon(x)\phi_1\Big(x, \frac{x}{\varepsilon}\Big)dx.\end{aligned}$$

After the passage to the limit $\varepsilon \to 0$ by employing the concept of the two-scale convergence, we obtain the problem (P_0), as follows.

Find the pair

$$(u_0, u_1) \in H_0^1(\Omega) \times L^2(\Omega; H_\#(Y)/\mathbb{R})$$

such that the identity

$$\frac{1}{|Y|}\int_\Omega\int_Y \chi(y)A(y)(\nabla u_0(x) + \nabla_y u_1(x,y))(\nabla\phi(x) + \nabla_y\phi_1(x,y))dxdy$$
$$= \frac{1}{|Y|}\int_\Omega\int_Y \chi(y)f_0(x,y)\phi_0(x)dxdy. \quad (4.15)$$

holds for all pairs of test functions

$$(\phi_0, \phi_1) \in H_0^1(\Omega) \times L^2(\Omega; H_\#(Y)/\mathbb{R}).$$

The structure of the obtained homogenized problem (P_0) is as one expects from the calculations done for the case involving homogeneous domains, as in the previous chapter. The main difference here is twofold:

(i) Due to the presence of the factor $\chi(y)$, the choice of the shape of the microstructure Y_0 affects the entries in the homogenized diffusivity tensor.
(ii) Depending on the choice of the boundary conditions imposed at Γ_ε, the homogenized problem might contain new terms, as we will see in Section 4.5.

We invite the reader to write explicitly the strong form of the limit problem, along with the corresponding effective coefficients.

4.4 Auxiliary Results Applicable to Perforated Domains

A number of specific theoretical results are required to perform the passage to the homogenization limit for partial differential equations posed on perforated domains. The purpose of this section is to present these results succinctly. We also provide a few brief proofs when doing so does not disrupt the flow of the presentation. With practical, real-world applications of periodic homogenization techniques in mind, our focus is exclusively on extension results in Sobolev spaces, Poincaré-type inequalities for perforated media, and trace inequalities for perforated media.

4.4.1 *Extension results*

We rely on extension results for Sobolev and Bochner spaces. Essentially, if we can show that the extension of u_ε is uniformly bounded in $H^1(\Omega)$, then we are done since we can simply use the two-scale convergence arguments as before.

(H5) Assume ∂Y_0 and $\partial\Omega_\varepsilon$ to be Lipschitz.

Lemma 4.1 (Extension lemma). *Assume* (H5). *The following statements hold:*

(i) (*Extension within the microscopic domain*) *If* $u \in H^1(Y)$, *then there exists* $\tilde{u}$ – *extension into* Y_0 (*and thus on* Z) *of* u – *such that*

$$\|\tilde{u}\|_{H^1(Z)} \leq c\|u\|_{H^1(Y)}.$$

(ii) (*Extension within the macroscopic domain*) *If* $u_\varepsilon \in H^1(\Omega_\varepsilon)$, *then there exists* $\tilde{u}_\varepsilon$ – *extension to* Ω *of* u_ε – *such that*

$$\|\tilde{u}_\varepsilon\|_{H^1(\Omega)} \leq c\|u_\varepsilon\|_{H^1(\Omega_\varepsilon)},$$

where $c > 0$ *is independent of* ε.

Proof. (i) We refer the reader to any textbook on function spaces where the topic extension in Sobolev spaces is treated; see e.g. Adams and Fournier (2003).

(ii) As a basic rule, we consider that the summation over $k \in \mathbb{Z}^d$ is such that $\varepsilon Y \subset \Omega$. We have

$$\begin{aligned}
\|\tilde{u}_\varepsilon\|^2_{H^1(\Omega)} &= \sum_{k\in\mathbb{Z}^d} \int_{\varepsilon Z^k} \left(|\tilde{u}_\varepsilon(x)|^2 + |\nabla\tilde{u}_\varepsilon(x)|^2\right)dx \\
&\overset{y=\frac{x}{\varepsilon}}{=} \sum_{k\in\mathbb{Z}^d} \varepsilon^d \int_{Z^k} \left(|\tilde{u}_\varepsilon(\varepsilon y)|^2 + \varepsilon^2|\nabla\tilde{u}_\varepsilon(\varepsilon y)|^2\right)dy \\
&\overset{extension}{\leq} c\sum_{k\in\mathbb{Z}^d} \varepsilon^d \int_{Y^k} \left(|u_\varepsilon(\varepsilon y)|^2 + \varepsilon^2|\nabla u_\varepsilon(\varepsilon y)|^2\right)dy \\
&= c\sum_{k\in\mathbb{Z}^d} \int_{\varepsilon Y^k} \left(|u_\varepsilon(x)|^2 + |\nabla u_\varepsilon(x)|^2\right)dx \\
&= c\|u_\varepsilon\|^2_{H^1(\Omega_\varepsilon)}.
\end{aligned}$$

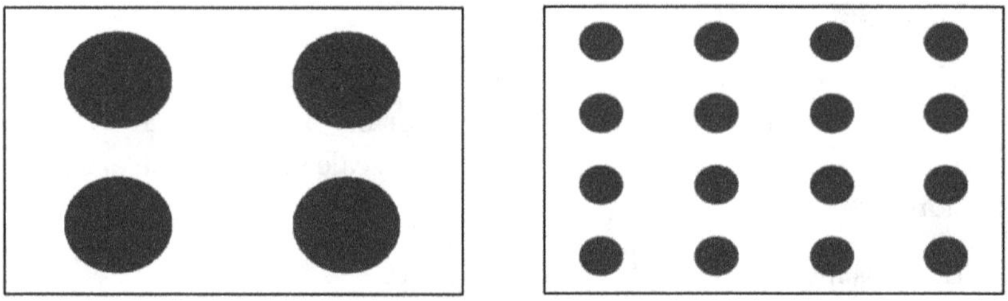

Figure 4.2. Sketch of two-dimensional perforated structures for two distinct choices of ε. The white and gray colors indicate two different materials.

Observe that the constant c arising in (i) and (ii) is the same. For more on this matter, see e.g. Hornung and Jäger (1991) (inspiration source for this variant of the proof) or Proposition 3.50 in Cioränescu and Donato (1999). □

The microstructures pointed out in Figure 4.2 allow for suitable extensions. As far as we are aware, the best extension results regarding homogenization scenarios are currently the ones reported by Acerbi *et al.* (1992), together with the refinements brought in by M. Höpker as detailed in his PhD thesis (Höpker, 2016). The reader might take the challenge of sketching with pencil and paper microstructures that fit to the geometric setting from Acerbi *et al.* (1992) and others that are are pathological, in the sense that they cannot be treated (for now) due to a lack of suitable extension results[3]; see such an example shown in Figure 4.3.

4.4.2 *Poincaré-type inequalities*

Lemma 4.2 (A Poincaré–Wirtinger-type inequality). *Let $Y, Z \subset \mathbb{R}^d$, Y be a connected, open set in $\mathbb{R}^d$, and $Y \subseteq Z$ such that $0 < |Y| \leq |Z|$, with ∂Y and ∂Z of class C^1. Let also $p \in [1, \infty]$. Then, there exists a constant c_P depending only on d, p, Y, and Z such that*

$$\left\| \phi - \fint_Y \phi dy \right\|_{L^p(Y)} \leq c_P \|\nabla \phi\|_{L^p(Y)} \tag{4.16}$$

for each $\phi \in W^{1,p}(Y)$.

[3] In this case, rigorous averaging statements can be made using, in a direct way, the concept of two-scale convergence.

Figure 4.3. Non-averageable structures: Note the white cusps and geometric singularities. Courtesy of S. A. Muntean for photo and Pischinger cake.

Proof. We follow the line of arguments from Evans (2015, Theorem 1, pp. 275–276). The proof goes by contradiction. Assuming (4.16) to be false results in the fact that there would exist, for all $k \in \mathbb{N}$, a function $\phi_k \in W^{1,p}(Y)$ satisfying

$$\left\| \phi_k - \fint_Y \phi_k(x)dx \right\|_{L^p(Y)} > k\|\nabla\phi_k\|_{L^p(Y)}. \tag{4.17}$$

For all $k \in \mathbb{N}$, we define the function

$$w_k(x) = \frac{\phi_k(x) - \fint_Y \phi_k(y)dy}{\|\phi_k - \fint_Y \phi_k(y)dy\|_{L^p(Y)}}. \tag{4.18}$$

It is easy to see that

$$\fint_Y w_k(x)dx = 0 \quad \text{and} \quad \|w_k\|_{L^p(Y)} = 1. \tag{4.19}$$

Applying now (4.17) to w_k (a renormalized version of ϕ_k), we get

$$\|\nabla w_k\|_{L^p(Y)} < \frac{1}{k} \quad \text{for all } k \in \mathbb{N}. \tag{4.20}$$

Since w_k is bounded in $W^{1,p}(Y)$ and, additionally, since the embedding $W^{1,p}(Y) \hookrightarrow L^p(Y)$ is compact via Theorem 7.5 (Rellich–Kondrachov theorem), we deduce that there exists at least a subsequence $(w_{k_j}) \subset (w_k)$

such that

$$w_{k_j} \to w \text{ strongly in } L^p(Y), \quad \text{as } j \to \infty.$$

Using this strong convergence to pass to the limit $k \to \infty$ in (4.19) yields $⨍_Y w(x)dx = 0$ and

$$\|w\|_{L^p(Y)} = 1. \tag{4.21}$$

Furthermore, (4.20) also implies that, for all $i \in \{1, \ldots, d\}$ and all $\varphi \in C_0^\infty(Y)$, we have

$$\begin{aligned}\int_Y w(x)\frac{\partial \varphi}{\partial x_i}(x)dx &= \lim_{j\to\infty}\int_Y w_{k_j}(x)\frac{\partial \varphi_{k_j}}{\partial x_i}(x)dx\\ &= -\lim_{j\to\infty}\int_Y \frac{\partial w_{k_j}}{\partial x_i}(x)\varphi(x)dx\\ &= 0.\end{aligned}$$

So, Du Bois Reymond's lemma, together with $w \in W^{1,p}(Y)$, leads to $\nabla w = 0$ a.e. in Y. Thus, w is constant, as Y is a connected set. On the other hand, combining the fact that w is a constant with the fact that $⨍_Y w(x)dx = 0$ implies the vanishing of w, that is, $w = 0$ a.e. in Y. From here, we deduce that $\|w\|_{L^p(Y)} = 0$, and hence, $\|w\|_{L^p(Y)} = 0$. This contradicts (4.21), and hence, the proof of the lemma is completed. □

If $Y = Z$, then Lemma 4.2 points out the usual Poincaré–Wirtinger inequality. A standard reference regarding Poincaré-type inequalities is Saloff-Coste (2002).

Lemma 4.3 (Poincaré's inequality for perforated media). *There exists a constant $c_P > 0$, independent of ε, such that*

$$\|u_\varepsilon\|_{L^2(\Omega_\varepsilon)} \le c_P\|\nabla u_\varepsilon\|_{L^2(\Omega_\varepsilon)} \tag{4.22}$$

for all $u_\varepsilon \in H^1(\Omega_\varepsilon; \partial\Omega)$.

Proof. We adapt the proof from Ciorănescu and Saint Jean Paulin (1998, Lemma 2.1, pp. 14–15)). The extension lemma ensures the existence of a linear continuous extension operator

$$\mathcal{P} : H^1(\Omega_\varepsilon; \partial\Omega) \to H_0^1(\Omega)$$

such that $\mathcal{P}u_\varepsilon = \tilde{u}_\varepsilon = u_\varepsilon$ a.e. on Ω_ε and

$$\|\nabla(\mathcal{P}u_\varepsilon)\|_{L^2(\Omega)} \leq c^*\|\nabla u_\varepsilon\|_{L^2(\Omega)}, \tag{4.23}$$

where c^* is the extension constant. Note that c^* is independent of the choice of ϵ. By the classical Poincaré's inequality, we have that

$$\|\mathcal{P}u_\varepsilon\|_{L^2(\Omega)} \leq \tilde{c}_P\|\nabla(\mathcal{P}u_\varepsilon)\|_{L^2(\Omega)}, \tag{4.24}$$

where, obviously, the positive constant $\tilde{c}_P$ does depend on Ω and d, but it is independent on ε. We conclude now the proof of this lemma by noting that

$$\begin{aligned}
\|u_\varepsilon\|_{L^2(\Omega_\varepsilon)} &\overset{\text{extension in the interior}}{\leq} c_\star\|\mathcal{P}u_\varepsilon\|_{L^2(\Omega)} \\
&\overset{\text{Poincaré's inequality}}{\leq} c_\star\tilde{c}_P\|\nabla(\mathcal{P}u_\varepsilon)\|_{L^2(\Omega)} \\
&\overset{\text{extension in the exterior}}{\leq} c_\star\tilde{c}_P c^\star\|\nabla u_\varepsilon\|_{L^2(\Omega_\varepsilon)}.
\end{aligned}$$

The last step is obtained using (4.23) and Lemma 4.1(ii). Take now

$$c_P = c_\star\tilde{c}_P c^\star > 0, \tag{4.25}$$

and note that c_P is independent of ε. □

4.4.3 *Trace inequalities*

The following result is a trace inequality tailored for perforated media. This is a very useful inequality for those applications where at least part of the boundary is active. By "active boundary", we mean that boundary processes (such as those introduced via Cauchy fluxes, cf. Gurtin (1993)) are defined there by means of boundary production terms. This is often described by the presence of a Robin-type boundary condition posed on a subset of Γ_ε.

Lemma 4.4 (Trace inequality). *Let $\varphi \in H^1(\Omega_\varepsilon)$. Then, there exists a constant $\hat{c} > 0$, independent of ε, such that the following inequality holds:*

$$\varepsilon\|\varphi\|^2_{L^2(\Gamma_\varepsilon)} \leq \hat{c}\|\varphi\|^2_{H^1(\Omega_\varepsilon)}. \tag{4.26}$$

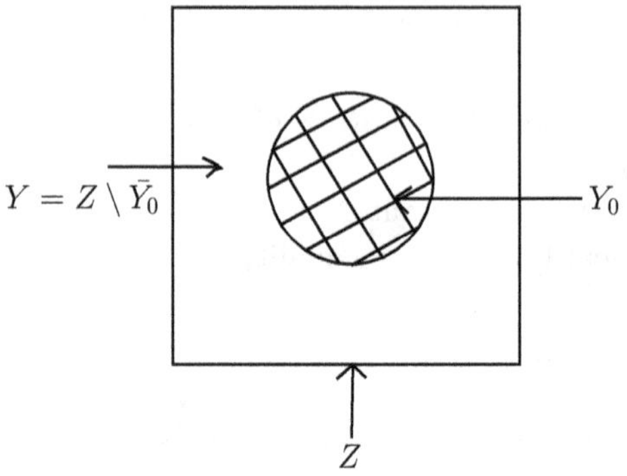

Figure 4.4. Basic geometry of the microstructure.

Proof. Consider Figure 4.4 as a reference microstructure image. To fix ideas, we set

$$Y = Z\backslash\bar{Y}_0 \quad \text{and} \quad \Gamma = \partial Y_0,$$

and let n be the outer normal vector on Γ with $||n|| = 1$, where $||\cdot||$ is here the Euclidean norm in $\mathbb{R}^d$. We introduce the smooth extension $N = (N_1, \ldots, N_d)$ of the normal unit vector n by

$$N_j(y) = n_j(y) \quad \text{for all } y \in \Gamma \tag{4.27}$$

for all $j \in \{1, \ldots, d\}$ such that supp N_j is included in a closed "tubular" neighborhood of Γ. We can also accommodate the restriction $N_j = 0$ at $\partial\Omega = \Gamma^{ext}$. In fact, we extend smoothly N by zero everywhere away from a neighborhood of Γ.

It is interesting to note that if $d \in \{2, 3\}$, then that tubular neighborhood of Γ resembles Gurtin's pillbox, which is sometimes used in continuum mechanics; see the pillbox lemma in Gurtin (1993).

Sufficient regularity of $\partial Y_0 = \Gamma$ is needed to support the following calculations.[4] We can therefore assume that we can choose $N_j \in C^1(\bar{Z})$. Due to the periodicity constraint and the fact that $N \| n$ (and so $(N, n)_{\sigma_\varepsilon} = 1$),

[4]Regarding the regularity of the boundary of the microstructure, we certainly do not need the interfaces to be infinitely smooth, like for balls, e.g. $\Gamma \in C^2$ is more than sufficient. Note however that this regularity constraint can be relaxed further if requested by a specific application.

we have that

$$\int_{\Gamma_\varepsilon} \varphi^2 d\sigma_\varepsilon = \int_{\Gamma_\varepsilon} \varphi^2 \sum_{j=1}^{d} N_j\left(\frac{x}{\varepsilon}\right) n_j\left(\frac{x}{\varepsilon}\right) d\sigma_\varepsilon = \int_{\Gamma_\varepsilon} \varphi^2(x) N\left(\frac{x}{\varepsilon}\right) n\left(\frac{x}{\varepsilon}\right) d\sigma_\varepsilon$$

$$= \int_{\Omega_\varepsilon} \operatorname{div}\left(\varphi^2(x) N\left(\frac{x}{\varepsilon}\right)\right) dx = \int_{\Omega_\varepsilon} \sum_{j=1}^{d} \frac{\partial}{\partial x_j}\left(\varphi^2(x) N_j\left(\frac{x}{\varepsilon}\right)\right) dx$$

$$= \sum_{j=1}^{d}\left[2\int_{\Omega_\varepsilon} N_j\left(\frac{x}{\varepsilon}\right)\varphi(x)\frac{\partial \varphi}{\partial x_j}(x)dx + \frac{1}{\varepsilon}\int_{\Omega_\varepsilon} \frac{\partial N_j\left(\frac{x}{\varepsilon}\right)}{\partial y_j}\varphi^2(x)dx\right].$$

Consequently, we obtain

$$\|\varphi\|^2_{\Gamma_\varepsilon} \leq \sum_{j=1}^{d}\left[2\max_{x\in\bar{\Omega}_\varepsilon}\left|N_j\left(\frac{x}{\varepsilon}\right)\right|\int_{\Omega_\varepsilon}|\varphi(x)\frac{\partial \varphi}{\partial x_j}(x)|dx\right]$$

$$+\sum_{j=1}^{d}\left[\frac{1}{\varepsilon}\max_{x\in\bar{\Omega}_\varepsilon}\left|\frac{\partial N_j\left(\frac{x}{\varepsilon}\right)}{\partial x_j}\right|\int_{\Omega_\varepsilon}\varphi^2(x)dx\right]$$

$$\leq c_1\|\varphi\|_{L^2(\Omega_\varepsilon)}\|\nabla\varphi\|_{L^2(\Omega_\varepsilon)} + \frac{1}{\varepsilon}c_2\|\varphi\|^2_{L^2(\Omega_\varepsilon)}.$$

We have taken here the constants

$$(c_1, c_2) := (2\|N\|_{L^\infty(\Omega_\varepsilon)}, \|\nabla N\|_{L^\infty(\Omega_\varepsilon)}),$$

where $N \in C^1(\bar{Z})$. We now have

$$\|\varphi\|^2_{L^2(\Gamma_\varepsilon)} \leq c_1\|\varphi\|_{L^2(\Omega_\varepsilon)}\|\nabla\varphi\|_{L^2(\Omega_\varepsilon)} + \frac{c_2}{\varepsilon}\|\varphi\|^2_{L^2(\Omega_\varepsilon)}.$$

We multiply the last inequality by ε. Using the geometric-arithmetic mean leads to

$$\varepsilon\|\varphi\|^2_{L^2(\Gamma_\varepsilon)} \leq \left(\frac{\varepsilon c_1^2}{2} + c_2\right)\|\varphi\|^2_{L^2(\Omega_\varepsilon)} + \frac{1}{2}\|\nabla\varphi\|^2_{L^2(\Omega_\varepsilon)} \leq \hat{c}\|\varphi\|^2_{H^1(\Omega_\varepsilon)}.$$

Note that $\hat{c} = \max\{c_1^2 + c_2, \frac{1}{2}\}$ is independent of the choice of ε. □

Another trace inequality useful for handling integral estimates for functions defined on oscillating hypersurfaces is introduced in the following lemma.

Lemma 4.5. *There exists $c > 0$ independent of ε such that for all $\varphi \in H^1(\Omega_\varepsilon)$ and for all $\delta > 0$, it holds*

$$\|\varphi\|^2_{L^2(\Gamma_\varepsilon)} \le c\left[\frac{1}{\varepsilon\delta}\|\varphi\|^2_{L^2(\Omega_\varepsilon)} + \varepsilon\delta\|\nabla\varphi\|^2_{L^2(\Omega_\varepsilon)}\right]. \tag{4.28}$$

Proof. This result is part of the homogenization folklore. We follow here the proof given by Ainouz, cf. Lemma 3.1 in Ainouz (2007).

We introduce the notation

$$\varphi^k_\varepsilon(y) := \varphi(\varepsilon(k+y)),$$

where $k \in K_\varepsilon$, with

$$K_\varepsilon := \{k \in \mathbb{Z}^d : \varepsilon(k+Y) \cap \Omega \neq \emptyset\}.$$

We have

$$\begin{aligned}\int_{\Gamma_\varepsilon} \varphi^2(x)d\sigma_\varepsilon(x) &= \sum_{k\in K_\varepsilon}\int_{\varepsilon(k+\Gamma)} \varphi^2(x)d\sigma_\varepsilon(x)\\ &= \varepsilon^{d-1}\sum_{k\in K_\varepsilon}\int_\Gamma \varphi^2(\varepsilon(k+y))d\sigma(y)\\ &= \varepsilon^{d-1}\sum_{k\in K_\varepsilon}\int_\Gamma [\varphi^k_\varepsilon(y)]^2 d\sigma(y).\end{aligned}$$

Using now the interpolation trace inequality (with $\nu = \frac{1}{2}$), we obtain

$$\begin{aligned}\int_\Gamma [\varphi^k_\varepsilon(y)]^2 d\sigma(y) &\le c\|\varphi^k_\varepsilon(x)\|^\nu_{H^1(\Omega_\varepsilon)}\|\varphi^k_\varepsilon(x)\|^{1-\nu}_{L^2(\Omega_\varepsilon)}\\ &\le \frac{c}{\varepsilon^d}\left[\frac{1}{\delta}\int_{\varepsilon(k+Y)} |\varphi(x)|^2dx + \delta\varepsilon^2\int_{\varepsilon(k+Y)} |\nabla\varphi(x)|^2dx\right].\end{aligned}$$

Hence, it holds that

$$\int_{\Gamma_\varepsilon} \varphi^2(x) d\sigma_\varepsilon(x) = \varepsilon^{d-1} \sum_{k\in K_\varepsilon} \int_\Gamma [\varphi_\varepsilon^k(y)]^2 d\sigma(y)$$

$$= \varepsilon^{d-1} \frac{c}{\varepsilon^d} \left[\frac{1}{\delta} \int_{\varepsilon(k+Y)} |\varphi(x)|^2 dx + (\delta\varepsilon)\varepsilon \int_{\varepsilon(k+Y)} |\nabla\varphi(x)|^2 dx \right]$$

$$\leq c \left[\frac{1}{\varepsilon\delta} \int_{\Omega_\varepsilon} |\varphi(x)|^2 dx + \varepsilon\delta \int_{\Omega_\varepsilon} |\nabla\varphi(x)|^2 dx \right],$$

which completes the proof of the statement. □

4.5 Averaging Boundary Processes in Perforated Media

We focus our attention on the following microscopic problem:

$$\operatorname{div}\big(- a_\varepsilon(x)\nabla u_\varepsilon(x)\big) = f_\varepsilon(x) \quad \text{for } x \in \Omega_\varepsilon, \tag{4.29}$$

$$-a_\varepsilon(x)\nabla u_\varepsilon(x) \cdot n_\varepsilon(x) = \varepsilon g_\varepsilon(x) \quad \text{for } x \in \Gamma_\varepsilon, \tag{4.30}$$

$$u_\varepsilon(x) = 0 \quad \text{for } x \in \partial\Omega. \tag{4.31}$$

We refer to problems (4.29)–(4.31) as $(\hat{P}_\varepsilon)$.

We assume (H1)–(H2), (H4), and (H5) to hold true and rephrase (H3) so that it states that we consider f_ε and g_ε to be uniformly bounded in $L^2(\Omega_\varepsilon)$ and $L^2(\Gamma_\varepsilon)$, that is:

(H3) there exists $c > 0$, which is independent of ε, so that

$$\|f_\varepsilon\|_{L^2(\Omega_\varepsilon)} + \varepsilon^{\frac{1}{2}} \|g_\varepsilon\|_{L^2(\Gamma_\varepsilon)} \leq c. \tag{4.32}$$

Here, we have in mind the structures $f_\varepsilon(x) = f_0\left(x, \frac{x}{\varepsilon}\right)$ for $x \in \Omega_\varepsilon$ and $g_\varepsilon(x) = g_0\left(x, \frac{x}{\varepsilon}\right)$ for $x \in \Gamma_\varepsilon$, where both functions $f_0(\cdot,\cdot)$ and $g_0(\cdot,\cdot)$ are sufficiently smooth with respect to their second variable.

It is convenient to introduce now the following Sobolev space:

$$H^1(\Omega_\varepsilon; \partial\Omega) = \{\varphi \in H^1(\Omega_\varepsilon) : \varphi = 0 \text{ at } \partial\Omega\}.$$

Definition 4.2 (Weak solution to $(\hat{P}_\varepsilon)$). The function $u_\varepsilon \in H^1(\Omega_\varepsilon; \partial\Omega)$ is called a weak solution to $(\hat{P}_\varepsilon)$ if and only if for all $\varphi \in H^1(\Omega_\varepsilon; \partial\Omega)$, the following identity holds:

$$\int_{\Omega_\varepsilon} a_\varepsilon(x)\nabla u_\varepsilon(x)\nabla\varphi(x)dx = \int_{\Omega_\varepsilon} f_\varepsilon(x)\varphi(x)dx + \varepsilon\int_{\Gamma_\varepsilon} g_\varepsilon(x)\varphi(x)d\sigma_\varepsilon. \tag{4.33}$$

A suitable application of the Lax–Milgram lemma ensures the existence and uniqueness of a $u_\varepsilon \in H^1(\Omega_\varepsilon; \partial\Omega)$ satisfying (4.33). We invite the reader to provide the details of the proof of the weak solvability of $(\hat{P}_\varepsilon)$.

To perform the homogenization procedure in this case, two crucial steps need to be made:

(i) Prove that there exists $c > 0$ independent of ε so that $\|u_\varepsilon\|_{H^1(\Omega_\varepsilon;\partial\Omega)} \leq c$.
(ii) Adapt the concept of two-scale convergence to functions defined on oscillating hypersurfaces.

The main ingredients for proving (i) include suitable Poincaré and trace-type inequalities adapted to perforated media, while for dealing with (ii) we need a compactness theorem associated with some sort of two-scale convergence concept that can be applied to oscillating surfaces.

By Lemma 4.1, we deduce that there exists an extension $\tilde{u}_\varepsilon \in H_0^1(\Omega)$ of $u_\varepsilon \in H^1(\Omega_\varepsilon; \partial\Omega)$. We also consider $\tilde{f}_\varepsilon \in L^2(\Omega)$ as the zero extension of $f_\varepsilon \in L^2(\Omega_\varepsilon)$, as well as $\tilde{a}_\varepsilon \in L^2(\mathbb{R}^d)$ as the zero extension of $a_\varepsilon \in L^\infty(\Omega_\varepsilon)$. In this case, (4.33) becomes

$$\int_\Omega \tilde{a}_\varepsilon(x)\nabla\tilde{u}_\varepsilon(x)\nabla\varphi(x)dx = \int_\Omega \tilde{f}_\varepsilon(x)\varphi(x)dx + \varepsilon\int_{\Gamma_\varepsilon} g_\varepsilon(x)\varphi(x)d\sigma_\varepsilon \tag{4.34}$$

for $\varphi \in H_0^1(\Omega)$. Now, choose $\varphi(x) = \phi_0(x) + \varepsilon\phi_1\left(x, \frac{x}{\varepsilon}\right)$ as the test function, where

$$(\phi_0, \phi_1) \in C_0^\infty(\Omega) \times C_0^\infty(\Omega; C_\#^\infty(Y)).$$

We can see that we can pass to the two-scale convergence limit in the first two terms of (4.34). This can be done, as one can make use of the uniform bounds

$$\|\tilde{u}_\varepsilon\|_{H_0^1(\Omega)} \leq c_1\|u_\varepsilon\|_{H^1(\Omega_\varepsilon;\partial\Omega)} \leq c_2$$

and of Theorem 7.7, which together ensure that there exist

$$(u_0, u_1) \in H_0^1(\Omega) \times L^2(\Omega; H_\#^1(Y)/\mathbb{R})$$

such that

$$\tilde{u}_\varepsilon \stackrel{2}{\rightharpoonup} u_0(x,y)$$

and

$$\nabla \tilde{u}_\varepsilon \stackrel{2}{\rightharpoonup} \nabla_x u_0(x) + \nabla_y u_1(x,y).$$

The third term in (4.34) has a particular structure that fits well to the concept of two-scale convergence on oscillating hypersurfaces. So, we use the convergence concept defined in Definition 7.5 to pass to the limit $\varepsilon \to 0$ in this last term. Essentially, we have that the sequence of functions (g_ε) in $L^2(\Gamma_\varepsilon)$ two-scale converges to a limit $g_0 \in L^2(\Omega \times \Gamma)$ if and only if for any $\varphi \in C_0^\infty(\Omega, C_\#^\infty(\Gamma))$, we have

$$\lim_{\varepsilon\to 0} \epsilon \int_{\Gamma_\varepsilon} g^\varepsilon(x)\varphi\Big(x,\frac{x}{\epsilon}\Big) d\sigma_\varepsilon = \frac{1}{|Y|}\int_\Omega \int_\Gamma g_0(x,y)\varphi(x,y)\, d\sigma_y dx. \tag{4.35}$$

The bound (4.32) indicated in (H3) ensures that the last limit is calculated correctly. Consequently, we obtain

$$\lim_{\varepsilon\to 0} \epsilon \int_{\Gamma_\varepsilon} g^\varepsilon(x)\left(\phi_0(x) + \varepsilon\phi_1\Big(x,\frac{x}{\epsilon}\Big)\right) d\sigma_\varepsilon = \frac{1}{|Y|}\int_\Omega \int_\Gamma g_0(x,y)\phi_0(x) d\sigma_y dx.$$

After the passage to the limit $\varepsilon \to 0$ by employing the concept of the two-scale convergence, we obtain (along subsequences) the weak formulation of the two-scale limit problem $(\hat{P}_0)$, as follows.

Find the pair

$$(u_0, u_1) \in H_0^1(\Omega) \times L^2(\Omega; H_\#(Y)/\mathbb{R})$$

such that the identity

$$\begin{aligned} &\frac{1}{|Y|}\int_\Omega \int_Y A(y)(\nabla_x u_0(x) + \nabla_y u_1(x,y))(\nabla_x \phi(x) + \nabla_y \phi_1(x,y))dxdy \\ &\quad = \frac{1}{|Y|}\int_\Omega \int_Y f_0(x,y)\phi_0(x)dxdy + \frac{1}{|Y|}\int_\Omega \int_\Gamma g_0(x,y)\phi_0(x)d\sigma_y dx. \end{aligned} \tag{4.36}$$

holds for all pairs of test functions

$$(\phi_0, \phi_1) \in H_0^1(\Omega) \times L^2(\Omega; H_\#(Y)/\mathbb{R}).$$

Comparing now the weak two-scale limit formulations (P_0) and ($\hat{P}_0$) (i.e. (4.15) vs. (4.36)), we see that in the last result, we have obtained an additional term on the right-hand side of the upscaled equation. This is a direct macroscopic effect of the presence in the microscopic problem of the production terms $\varepsilon g_\varepsilon(\cdot)$ posed across the microscopic interfaces Γ_ε.

Here as well, we encourage the reader to write down explicitly the strong form of the limit problem, along with the corresponding effective coefficients.

4.6 Quest for Corrector Estimates

In the previous sections of this chapter, we studied the passage to the homogenization limit for two linear boundary-value problems posed for perforated domains. Specifically, we "homogenized" the boundary-value problems (4.1)–(4.2) and (4.29)–(4.31) by suitably employing the concept of two-scale convergence. As a consequence of this way of working, we discovered the limit function u_0, as well as limit equations that this satisfies. However, two closely related questions are still in need of answers:

> What is the error we make when we replace u_ε with u_0 for fixed but small values of the parameter $\varepsilon > 0$?

and

> How good or bad is the produced averaging? In other words, how far are the upscaled equations from the original microscopic problem when compared for a fixed small value of $\varepsilon > 0$?

Regarding problems (4.29)–(4.31), the following result gives the desired answer to the question concerning the quality of the averaging procedure.

Theorem 4.1. *Let u_ε be a weak solution to* (4.29)–(4.31). *Then, the following corrector estimate holds:*

$$||u_\varepsilon - u_0 - m_\varepsilon(\varepsilon u_1 + \varepsilon^2 u_2)||_{H^1(\Omega_\varepsilon)} \leq c\sqrt{\varepsilon}, \tag{4.37}$$

where m_ε is a suitable cutoff around $\partial\Omega$ and $c > 0$ is a constant independent of the choice of ε.

The proof of this result can be found in the work of Cioranescu and Saint Jean Paulin (1998). The calculations are rather lengthy. Therefore, we omit to show them here. However, it is one of the important proofs in this research field, and hence, a careful reader should check the original source. The proof uses all the detailed information one can obtain from the structure of the formal two-scale expansion applied to (4.29), together with at least two families of cell functions: one expressing u_1 in terms of $\nabla_x u_0$, and the other expressing u_2 in terms of $\Delta_x u_0$. It is quite instructive, as it offers a general methodology for proving corrector estimates for linear elliptic, parabolic, and pseudo-parabolic problems. As a direct corollary, one can also obtain

$$||u_\varepsilon - u_0||_{L^2(\Omega_\varepsilon)} \leq c\varepsilon. \tag{4.38}$$

Proving corrector estimates for a simple setting. Let us now justify the homogenization asymptotics for a simpler case. Our reference material for this paragraph is Bensoussan *et al.* (1978). We begin with assuming not only that the data (i.e. a_ε, f, and $\partial\Omega$) is sufficiently smooth but also that we have available to us the structure of the homogenization *ansatz*.

Let u_ε be a solution of problem (P_ε),

$$\operatorname{div}\big(- a_\varepsilon(x)\nabla u_\varepsilon(x)\big) = f(x) \quad \text{for } x \in \Omega, \tag{4.39}$$

$$u_\varepsilon(x) = 0 \quad \text{for } x \in \partial\Omega, \tag{4.40}$$

and u_0 be a solution of problem (P_0),

$$-\operatorname{div}\big(\mathbb{D}\nabla u_0(x)\big) = f(x) \quad \text{for } x \in \Omega, \tag{4.41}$$

$$u_0(x) = 0 \quad \text{for } x \in \partial\Omega. \tag{4.42}$$

Recall the form of the two-scale asymptotic expansion:

$$u_\varepsilon(x) = u_0(x) + \varepsilon u_1(x,y) + \varepsilon^2 u_2(x,y) + \mathcal{O}(\varepsilon^3), \tag{4.43}$$

where $y = \frac{x}{\varepsilon}$ and $u_1, u_2, \ldots$ are Y-periodic functions. Furthermore, we can express u_1 as $u_1(x,y) = W(y) \cdot \nabla_x u_0(x) + \tilde{u}_1(x)$, where $W = W(y)$ is the vector of cell functions, while $\tilde{u}_1(x)$ is some arbitrary function constant with respect to the variable y. Let us choose $\tilde{u}_1(x) = 0$, as it carries no importance in the following calculations.

The expansion (4.43) indicates a candidate for the expected error the following expression:

$$z_\varepsilon := u_\varepsilon - (u_0 + \varepsilon u_1 + \varepsilon^2 \mathcal{N}). \tag{4.44}$$

The key idea is to choose the function $\mathcal{N} = \mathcal{N}(x, y)$ so that the quantity $T_\varepsilon z_\varepsilon = \operatorname{div}(-a_\varepsilon(x)\nabla z_\varepsilon(x))$ is as small as possible. Recalling the structure of two-scale formal asymptotics, we have that

$$T_\varepsilon = \varepsilon^{-2}T_{-2} + \varepsilon^{-1}T_{-1} + \varepsilon^0 T_0.$$

We will see in the forthcoming calculations that an excellent candidate for N is precisely u_2. With this choice at hand, we claim that

$$T_\varepsilon z_\varepsilon = -\varepsilon r_\varepsilon, \tag{4.45}$$

where $r_\varepsilon = T_{-1}\mathcal{N} + T_0 u_1 + \varepsilon T_0 \mathcal{N}$. The operators T_{-2}, T_{-1}, and T_0 have been introduced previously in Section 2.3. For convenience, we recall their definition here:

$$T_{-2}(\cdot) = -\sum_{i=1}^{d}\sum_{j=1}^{d} \frac{\partial}{\partial y_i}\left(a_{ij}(y)\frac{\partial}{\partial y_j}\right); \tag{4.46}$$

$$T_{-1}(\cdot) = -\sum_{i=1}^{d}\sum_{j=1}^{d} \left[\frac{\partial}{\partial y_i}\left(a_{ij}(y)\frac{\partial}{\partial x_j}\right) + \frac{\partial}{\partial x_i}\left(a_{ij}(y)\frac{\partial}{\partial y_j}\right)\right]; \tag{4.47}$$

$$T_0(\cdot) = -\sum_{i=1}^{d}\sum_{j=1}^{d} \frac{\partial}{\partial y_i}\left(a_{ij}(y)\frac{\partial}{\partial x_j}\right). \tag{4.48}$$

It is crucial to observe that the following hierarchy of equations holds:

$$T_{-2}u_0 = 0 \tag{4.49}$$

$$T_{-2}u_1 + T_{-1}u_0 = 0 \tag{4.50}$$

$$T_{-2}u_2 + T_{-1}u_1 + T_0 u_0 = f \tag{4.51}$$

$$T_{-2}u_3 + T_{-1}u_2 + T_0 u_1 = 0 \tag{4.52}$$

$$T_{-2}u_4 + T_{-1}u_3 + T_0 u_2 = 0 \ldots \tag{4.53}$$

The first three equations have been obtained by us while performing the two-scale formal homogenization asymptotics. Equations (4.52), (4.53) and so on represent the so-called high-order asymptotics. They are obtained by repeating the formal homogenization arguments in a suitable manner. The procedure is an iterative one. Knowing the structure of high-order asymptotics plays a crucial role in finding the corrector estimates or convergence rates for the homogenization limiting procedure; see e.g. Ciorănescu and Saint Jean Paulin (1998).

Now, we can prove our claim (4.45). By direct calculations and using the hierarchy of equations mentioned previously, we obtain

$$\begin{aligned}
T_\varepsilon z_\varepsilon &= (\varepsilon^{-2}T_{-2} + \varepsilon^{-1}T_{-1} + \varepsilon^{0}T_0)(u_\varepsilon - (u_0 + \varepsilon u_1 + \varepsilon^2 u_2))\\
&= T_\varepsilon u_\varepsilon - (T_{-2}u_2 + T_{-1}u_1 + T_0 u_0)\\
&\quad - \varepsilon^{-2}T_{-2}u_0 - \varepsilon^{-1}(T_{-2}u_1 + T_{-1}u_0)\\
&\quad - (T_0 u_1 + T_{-1}u_2 + \varepsilon T_0 u_2)\\
&= f - f - \varepsilon^{-2}0 + \varepsilon^{-1}0 - \varepsilon r_\varepsilon.
\end{aligned}$$

Consequently, we can now study the following elliptic boundary-value problem: Find z_ε such that the following two equations are satisfied:

$$T_\varepsilon z_\varepsilon = -\varepsilon r_\varepsilon \quad \text{in } \Omega, \tag{4.54}$$

$$z_\varepsilon = -(\varepsilon u_1 + \varepsilon^2 u_2) \quad \text{at } \partial\Omega. \tag{4.55}$$

If the data of the problem is smooth, then u_ε and u_0 are smooth as well.[5] In this case, we can show that there exist two constants $c_1 > 0$ and $c_2 > 0$, independent of the choice of ε, such that the following estimates hold:

$$||r_\varepsilon||_{L^\infty(\Omega)} \leq c_1, \tag{4.56}$$

and

$$||z_\varepsilon||_{L^\infty(\partial\Omega)} \leq c_2\varepsilon. \tag{4.57}$$

Applying now the classical maximum principle for elliptic partial differential equations (cf. e.g. Theorem 4.25 in Renardy and Rogers (2004)) to the boundary-value problems (4.54)–(4.55) yields the existence of $c_3 > 0$ such

[5]The reader may want to find out how much regularity is needed on the data to ensure that $u_\varepsilon \in C^2(\Omega) \cap C(\bar{\Omega})$.

that

$$||z_\varepsilon||_{L^\infty(\Omega)} \leq c_3\varepsilon. \tag{4.58}$$

Furthermore, we also observe that

$$\begin{aligned}||u_\varepsilon - u_0||_{L^\infty(\Omega)} &= ||z_\varepsilon + \varepsilon u_1 + \varepsilon^2 u_2||_{L^\infty(\Omega)}\\ &\leq ||z_\varepsilon||_{L^\infty(\Omega)} + \varepsilon||u_1 + \varepsilon u_2||_{L^\infty(\Omega)}.\end{aligned}$$

We can now conclude the discussion by stating the following result:

Theorem 4.2. *Under suitable smoothness assumptions on data, there exists $c_4 > 0$ (independent of ε) such that*

$$||u_\varepsilon - u_0||_{L^\infty(\Omega)} \leq c_4\varepsilon. \tag{4.59}$$

The estimate (4.59) holds true if one replaces the space $L^\infty(\Omega)$ with any $L^p(\Omega)$ with $p \geq 1$. It is worth noting that the approximation $u_0(x) - \varepsilon u_1(x, \frac{x}{\varepsilon})$ cannot satisfy an homogeneous Dirichlet condition. Relying on the fact that $u_0 = 0$ at $\partial\Omega$, we note that $u_1(x, \frac{x}{\varepsilon})$ would need to vanish for all $x \in \partial\Omega$, but there are no reasons for this to happen. To correct for an eventual mismatch, one could introduce a function θ_ε satisfying the boundary-value problem

$$\begin{aligned}\operatorname{div}(-a_\varepsilon(x)\nabla\theta_\varepsilon(x)) = 0 \quad &\text{for } x \in \Omega\\ \theta_\varepsilon(x) = u_1\left(x, \frac{x}{\varepsilon}\right) \quad &\text{for } x \in \partial\Omega.\end{aligned}$$

The function θ_ε is usually referred to as a boundary corrector, as it enforces the relation

$$u_\varepsilon(x) - u_0(x) - \varepsilon\left(u_1\left(x, \frac{x}{\varepsilon}\right) - \theta_\varepsilon(x)\right) = 0 \quad \text{for all } x \in \partial\Omega.$$

Suppose now that we can prove that $u_0 \in H^2(\Omega)$. Then, based on Theorem 6.4 in Alouges (2016), we can get a constant $c > 0$ independent of ε so that the following improved corrector estimate holds:

$$\left|\left|u_\varepsilon(x) - u_0(x) - \varepsilon\left(u_1\left(x, \frac{x}{\varepsilon}\right) - \theta_\varepsilon(x)\right)\right|\right|_{H^1(\Omega)} \leq \varepsilon||u_0||_{H^2(\Omega)}. \tag{4.60}$$

Observe that (4.60) is an improved rate of convergence.

There are a couple of ways to prove corrector estimates. The most intuitive ones combine ideas from the use of the formal two-scale asymptotics within the frame of the energy method. Fourier analysis and Green functions-based estimates can turn out to be quite useful, especially for linear problems; see Aleksanyan *et al.* (2013). Other working ideas involve the periodic unfolding operator, eventually combined with FEM-like interpolations, as given by Griso (2004). The main ingredients needed to ensure success in proving a corrector-type estimate are the regularity of the solutions and the precise structure of the problem (nonlinearities, coupling). There are also examples of situations in homogenization theory where proving correctors is an open problem. This is the case, for instance, when handling multiple-porosity materials, stochastic homogenization, or homogenization of singular measures and when rough boundaries are part of the problem.

4.7 How to Bring ε in Model Equations?

Even for cases where it is obvious that periodic homogenization techniques are applicable and can contribute *de facto* to a better understanding of things, the precise meaning of the small parameter ε is not obvious. Sometimes, there could even be several alternative ways to introduce ε for a given application. It is the role of the modeler to decide not only on what equations should one consider but also to bring ε in the model equations. To decide on whether to have or not to have ε in the coefficients or in the model equations requires special attention. Normally, two distinct steps are involved:

- non-dimensionalization,
- scaling.

The interested reader can find more on this subject, for instance, the works of Barenblatt (2003) and Muntean (2015) and references cited therein. To provide here the full details will force us to deviate from the general purpose of this textbook. Instead, we propose a "learning-by-doing" approach guided by exercises.

The reader should note that there are various alternative ways to introduce ε into model equations. We find the strategy outlined by Knoch *et al.* (2023) to be particularly well-explained. To keep things somewhat mysterious, we would like to emphasize that, regardless of the choice of scaling in terms of ε, it remains challenging to determine whether the model is correct. Further investigations are typically required to draw conclusions on this matter.

4.8 Exercises

Exercise 4.8.1. Why is the constant $\frac{c_P}{\alpha}$ entering (4.10) independent of the choice of ε? Justify your answer.

Exercise 4.8.2. Prove that the constant $\tilde{c}$ from the inequality (4.11) is independent of ε.

Exercise 4.8.3. Justify why the zero extension $\tilde{u}_\varepsilon(x) = \chi(\frac{x}{\varepsilon})u_\varepsilon(x)$ satisfies the microscopic problem posed by Ainouz (2007).

Exercise 4.8.4. Obtain rigorously (4.15). Derive from here the corresponding strong formulation of $(P0)$. How do the cell problems look like?

Exercise 4.8.5. Show how does the constant c_P entering (4.25) depend on ε. Justify your answer.

Exercise 4.8.6. Pass to $\varepsilon \to 0$ in (4.34) and derive the limit two-scale equation.

Exercise 4.8.7. Let Ω be a bounded non-empty domain in $\mathbb{R}^d$, and assume an array of ε-periodic distributed perforations placed in Ω to form the set Ω_ε. The notation is as in the lecture. Consider the following problem, say (P_ϵ): Find the couple $(u_\varepsilon, v_\varepsilon)$ satisfying the model equations

$$\operatorname{div}(-a_\varepsilon(x)\nabla u_\varepsilon(x)) = 0 \quad \text{for } x \in \Omega_\varepsilon,$$

$$\operatorname{div}(-b_\varepsilon(x)\nabla v_\varepsilon(x)) = 0 \quad \text{for } x \in \Omega_\varepsilon,$$

$$-a_\varepsilon(x)\nabla u_\varepsilon(x)\cdot n_\varepsilon(x) = \varepsilon k_\varepsilon(x)(-f_\varepsilon(x) + Hv_\varepsilon(x)) \quad \text{for } x \in \Gamma_\varepsilon,$$

$$-b_\varepsilon(x)\nabla v_\varepsilon(x)\cdot n_\varepsilon(x) = -\varepsilon k_\varepsilon(x)(-f_\varepsilon(x) + Hv_\varepsilon(x)) \quad \text{for } x \in \Gamma_\varepsilon,$$

$$u_\varepsilon(x) = 0 = v_\varepsilon(x) \quad \text{for } x \in \partial\Omega.$$

Assume $H > 0$, $k_\varepsilon \in L^\infty_+(\Gamma_\varepsilon)$, $f_\varepsilon \in L^2(\Gamma_\varepsilon)$, where $f_\varepsilon(x) = f(x, \frac{x}{\varepsilon}) := f_0(x) + \varepsilon^3 f_1(x, \frac{x}{\varepsilon})$ (with f_0 and f_1 of order of $\mathcal{O}(1)$) and $k_\varepsilon(x) := k(\frac{x}{\varepsilon})$, $a_\varepsilon(x) := A(\frac{x}{\varepsilon})$ and $b_\varepsilon(x) := B(\frac{x}{\varepsilon})$, with A, B being $d \times d$-matrices and $x \in \Omega$. Furthermore, $n_\varepsilon(x) = n(\frac{x}{\varepsilon})$ is the normal vector to Γ_ε, while all functions are assumed to be Y-periodic in the variable x/ε.

(i) Make meaningful assumptions on the data and parameters of the problem so that the Lax–Milgram lemma ensures weak solutions in $H^1(\Omega_\varepsilon; \partial\Omega) \times H^1(\Omega_\varepsilon; \partial\Omega)$.

(ii) Perform the homogenization asymptotics $\varepsilon \to 0$ using the concept of two-scale convergence. Derive explicitly the strong formulation of the limit (macroscopic) system.

(iii) Redo the homogenization limit (via two-scale convergence arguments) for the case when $H = \varepsilon$.

(iv) Redo the homogenization limit (via two-scale convergence arguments) for the case when one replaces the given function f_ε by the unknown function $f_\varepsilon + u_\varepsilon$.

(v) If one replaces f_ε only with u_ε, then the resulting microscopic problem admits zero as a unique solution. If one takes the same microscopic model but instead of the current right-hand side "RHS", one takes "-RHS", then the working assumptions under which the weak solvability (and hence also the homogenization asymptotics) can be done are more relaxed. We invite the reader to spot which assumptions can be removed in this case.

Exercise 4.8.8. We consider the following mixed Dirichlet–Robin problem. Let Ω be a bounded domain in $\mathbb{R}^d$ with Lipschitz $\partial\Omega$. Take the perforated domain $\Omega_\varepsilon \subset \Omega$ to be connected such that the perforations do not touch each other and, additionally, that $\Gamma_\varepsilon = \Gamma_\varepsilon^D \cup \Gamma_\varepsilon^N$ is Lipschitz, with $\Gamma_\varepsilon^D \cap \Gamma_\varepsilon^N = \emptyset$ and $\lambda^{d-1}(\Gamma_\varepsilon^D) \neq 0$; $n_\varepsilon(x) = n(\frac{x}{\varepsilon})$ is the normal vector to Γ_ε^N. Consider the following boundary-value problem, say (P_ε):

$$\operatorname{div}\big(-a_\varepsilon(x)\nabla u_\varepsilon(x)\big) = -\sqrt{\varepsilon}u_\varepsilon(x) + f_\varepsilon(x) \quad \text{for } x \in \Omega_\varepsilon,$$
$$-a_\varepsilon(x)\nabla u_\varepsilon(x)\cdot n_\varepsilon(x) = \varepsilon u_\varepsilon(x) \quad \text{for } \Gamma_\epsilon^N,$$
$$u_\varepsilon(x) = 0 \quad \text{for } x \in \Gamma_\varepsilon^D \cup \partial\Omega.$$

Assume f_ε to be bounded uniformly with respect to ε in $L^2(\Omega_\varepsilon)$, and let a_ε satisfy assumptions (H1)–(H4); $n_\varepsilon(x) = n(\frac{x}{\varepsilon})$ is the normal vector to Γ_ε^N.

(i) Show the existence and uniqueness of a weak solution u_ε to (P_ε).

(ii) Prove that u_ε is uniformly bounded in $H^1(\Omega_\varepsilon; \Gamma_\varepsilon^D \cup \partial\Omega)$.

(iii) Show that there exists at least an extension of u_ε that is bounded in $H_0^1(\Omega)$.

(iv) Pass to the limit $\epsilon \to 0$ in (P_ε) via two-scale convergence and determine the weak form of the two-scale limit problem, say (P_0).

(v) Prove the uniqueness of the weak solutions to (P_0).
(vi) Obtain the strong formulation of (P_0). Eliminate u_1.
(vii) Use the expansion $u_\varepsilon(x) = u_0(x, \frac{x}{\varepsilon}) + \varepsilon u_1(x, \frac{x}{\varepsilon}) + \mathcal{O}(\varepsilon^2)$ and pass to the limit $\varepsilon \to 0$ in (P_ε) via the formal asymptotic homogenization procedure. Compare your result with the answer to the previous point.
(viii) Solve numerically the microscopic problem (P_ε) and the corresponding macroscopic problem (P_0). How do the solutions compare? Illustrate how the solutions to the cell problems look like.

Exercise 4.8.9. Under which additional assumptions can one prove the corrector estimate

$$\|u_\varepsilon - u_0\|_{H^1(\Omega_\varepsilon)} \leq c\varepsilon^{\frac{1}{2}},$$

where the constant $c > 0$ is independent of the parameter ε? Here, u_ε satisfies the first boundary value problem discussed in this chapter and u_0 is the solution of the corresponding homogenized equation.

Exercise 4.8.10. Let Ω be a bounded domain in $\mathbb{R}^d$ ($d \in \{2, 3\}$) with Lipschitz boundary $\partial\Omega$. Take the perforated domain $\Omega_\varepsilon \subset \Omega$ to be connected such that the perforations do not touch each other and that the boundary of the perforations Γ_ε is Lipschitz. Additionally, take $\Gamma_\varepsilon = \Gamma_\varepsilon^D \cup \Gamma_\varepsilon^N$, where $\Gamma_\varepsilon^D \cap \Gamma_\varepsilon^N = \emptyset$ and $\lambda^{d-1}(\Gamma_\varepsilon^D) \neq 0$. Γ_ε^D is the Dirichlet boundary, while Γ_ε^N is the corresponding Neumann boundary. Introduce the Hilbert space

$$V_\varepsilon := \{v \in H^1(\Omega_\varepsilon) : v = 0 \text{ at } \partial\Omega \cup \Gamma_\varepsilon^D\}.$$

Consider the following problem, say (P_ε): Find $u_\varepsilon \in V_\varepsilon$ such that it holds:

$$\operatorname{div}\big(-a_\varepsilon(x)\nabla u_\varepsilon(x)\big) = -u_\varepsilon(x) + f(\varepsilon|Y_0|) \quad \text{for } x \in \Omega_\varepsilon,$$
$$-a_\varepsilon(x)\nabla u_\varepsilon(x) \cdot n_\varepsilon(x) = \varepsilon g_\varepsilon(x) \quad \text{for } x \in \Gamma_\varepsilon^N,$$
$$u_\varepsilon(x) = 0 \quad \text{for } x \in \partial\Omega \cup \Gamma_\varepsilon^D.$$

Assume $f \in C(\mathbb{R})$ with $f(0) = 0$ and g_ε to be bounded in $L^2(\Gamma_\varepsilon)$ uniformly with respect to ε. In this context, $n_\varepsilon(x) = n(\frac{x}{\varepsilon})$ is the normal vector to Γ_ε^N, the matrix $a_\varepsilon(x) = A(\frac{x}{\varepsilon})$ and the scalar $g_\varepsilon(x) = g(\frac{x}{\varepsilon})$ are 1-periodic functions, while $|Y_0|$ denotes the volume of the non-empty d-dimensional set Y_0 (the void).

(i) Specify the conditions needed for a_ε so that the Lax–Milgram lemma ensures the existence and uniqueness of a weak solution u_ε to (P_ε). Justify your statement.
(ii) Prove that u_ε is uniformly bounded in V_ε.
(iii) Show that $u_\varepsilon \in V_\varepsilon$ can be extended to $\tilde{u}_\varepsilon$, which is actually bounded in $H^1(\Omega)$.
(iv) Pass to the limit $\epsilon \to 0$ in (P_ε) via two-scale convergence, and determine the weak form of the two-scale limit problem, say (P_0), with solution u_0.
(v) Prove the existence and uniqueness of the weak solution to (P_0).
(vi) Obtain the strong formulation of (P_0). Eliminate u_1.
(vii) Argue in two different ways why $u_0 \in L^1(\Omega)$, i.e. the system contains "finite mass".
(viii) Now, assume that perforations can touch the boundary of the periodic cell and/or they intersect the outer boundary $\partial\Omega$. Can you identify additional restrictions on the data and parameters so that the homogenization process can still be done via two-scale convergence arguments? Justify your answer.

Exercise 4.8.11. Let Ω be a bounded domain in $\mathbb{R}^2$ with Lipschitz $\partial\Omega$. Take the perforated domain $\Omega_\varepsilon \subset \Omega$ to be connected such that the perforations do not touch each other, and assume additionally that Γ_ε is Lipschitz. Consider the following problem, say (P_ε):

$$\operatorname{div}\big(-a_\varepsilon(x)\nabla u_\varepsilon(x)\big) + k_\varepsilon(x)u_\varepsilon(x) = v_\varepsilon(x) \quad \text{for } x \in \Omega_\varepsilon,$$

$$-\operatorname{div}(\varepsilon^2 b_\varepsilon(x)\nabla v_\varepsilon(x)) = u_\varepsilon(x) \quad \text{for } x \in \Omega_\varepsilon,$$

$$-a_\varepsilon(x)\nabla u_\varepsilon(x)\cdot n_\varepsilon(x) = \varepsilon b_\varepsilon(x)\nabla v_\varepsilon(x)\cdot n_\varepsilon(x) \quad \text{for } x \in \Gamma_\varepsilon,$$

$$u_\varepsilon(x) = v_\varepsilon(x) = 0 \quad \text{for } x \in \partial\Omega.$$

Assume f_ε and g_ε to be bounded (uniformly with respect to ε) in $L^2(\Omega_\varepsilon)$ and, respectively, in $L^2(\Gamma_\varepsilon)$, sufficiently smooth in their second variable. Additionally, let $k_\varepsilon \in L^\infty_\#(Y)$ and $a_\varepsilon, b_\varepsilon$ satisfy assumptions (H1)–(H4); $n_\varepsilon(x) = n(\frac{x}{\varepsilon})$ is the normal vector to Γ_ε.

(i) Show the existence and uniqueness of a weak solution $(u_\varepsilon, v_\varepsilon)$ to (P_ε).
(ii) Prove that u_ε is uniformly bounded in $H^1_0(\Omega_\varepsilon)$. What happens with v_ε?
(iii) Show that u_ε is actually bounded in $H^1_0(\Omega)$. What happens with the extension of v_ε?
(iv) Pass in (P_ε) via two-scale convergence to the limit $\epsilon \to 0$, and determine the weak form of the two-scale limit problem, say (P_0).

(v) Prove the uniqueness of the weak solutions to (P_0).
(vi) Eliminate u_1, and obtain the strong formulation of (P_0) only in terms of u_0.

Exercise 4.8.12. Assume the same geometry and restrictions on data as in the previous exercise. Show that the function u_ε, which solves weakly the boundary-value problem

$$\operatorname{div}\big(-a_\varepsilon(x)\nabla u_\varepsilon(x)\big) = f_\varepsilon(x) \quad \text{for } x \in \Omega_\varepsilon,$$
$$u_\varepsilon(x) = 0 \quad \text{for } x \in \Gamma_\varepsilon \cup \partial\Omega,$$

converges to zero as $\varepsilon \to 0$.

Exercise 4.8.13. Let Ω be a bounded domain in $\mathbb{R}^d$ $(d \in \{2,3\})$ with Lipschitz boundary $\partial\Omega$ and $\alpha \in [0,\infty)$. For a fixed $\varepsilon \in (0,+\infty)$, take a periodically perforated domain $\Omega_\varepsilon \subset \Omega$ to be connected such that the perforations do not touch each other and that the boundary of the perforations Γ_ε is Lipschitz. Define conveniently a Hilbert space V_ε where you wish that your concept of solution makes sense. Take $n_\varepsilon(x) = n(\frac{x}{\varepsilon})$ to be the normal vector to Γ_ε.

Consider the following microscopic problem, say (P_ε): Find $u_\varepsilon \in V_\varepsilon$ such that it holds

$$\operatorname{div}\big(-a_\varepsilon(x)\nabla u_\varepsilon(x)\big) = -\alpha u_\varepsilon(x) \quad \text{for } x \in \Omega_\varepsilon,$$
$$-a_\varepsilon(x)\nabla u_\varepsilon(x)\cdot n_\varepsilon(x) = \varepsilon u_\varepsilon(x) \quad \text{for } x \in \Gamma_\varepsilon,$$
$$u_\varepsilon(x) = \varepsilon^3 h_\varepsilon(x) \quad \text{for } x \in \partial\Omega.$$

In this context, the matrix $a_\varepsilon(x) = A(\frac{x}{\varepsilon})$ and the scalar $h_\varepsilon(x) = h(\frac{x}{\varepsilon})$ are 1-periodic functions, while $|Y_0|$ denotes the volume of the non-empty d-dimensional set Y_0 (the perforation). Assume that the periodic extension of h (call it again h) has the property $h|_{\Gamma_\varepsilon} = 0$.

(i) Specify the additional conditions needed for $\alpha, a_\varepsilon, h_\varepsilon$ so that the Lax–Milgram lemma can ensure the existence and uniqueness of a weak solution u_ε to (P_ε). Justify your statement.
(ii) Prove that u_ε is uniformly bounded in V_ε. Is u_ε bounded also in $L^1(\Omega_\varepsilon)$? Can one have $u_\varepsilon \in L^1(\Gamma_\varepsilon)$? Justify your answers.
(iii) Extend $u_\varepsilon \in V_\varepsilon$ up to a function $\tilde{u}_\varepsilon$ bounded in (a subspace of) $H_0^1(\Omega)$.
(iv) Pass in (P_ε) via two-scale convergence to the limit $\varepsilon \to 0$, and determine the weak form of the two-scale limit problem, say (P_0), with

solution u_0. Do you need additional assumptions on $h_\varepsilon(\cdot)$ to be able to perform this step?

(v) Is the resulting effective diffusion coefficient (matrix) coercive? Is it positive definite? Justify your answers.

(vi) Prove the existence and uniqueness of the weak solution to (P_0).

(vii) Eliminate u_1. Obtain the strong formulation of (P_0).

(viii) Show that $u_0 \in L^1(\Omega)$. Can you prove this result in two conceptually different ways?

Exercise 4.8.14. Let Ω be a bounded domain in $\mathbb{R}^d$ $(d \in \{2,3\})$ with Lipschitz boundary $\partial\Omega$. For a fixed $\varepsilon \in (0,+\infty)$, take a perforated domain $\Omega_\varepsilon \subset \Omega$ to be connected such that the perforations do not touch each other and that the boundary of the perforations Γ_ε is Lipschitz. Here, $\partial\Omega_\varepsilon = \partial\Omega \cup \Gamma_\varepsilon$. Imagine Ω_ε to be a large square perforated by an array of periodically placed funny little patterns. Draw your domain Ω_ε.

Consider the following problem, say (P_ε): Find the pair $(u_\varepsilon, v_\varepsilon) \in H_0^1(\Omega_\varepsilon) \times H_0^1(\Omega_\varepsilon)$ satisfying

$$\operatorname{div}(-a_\varepsilon(x)\nabla u_\varepsilon(x)) = v_\varepsilon(x) - u_\varepsilon(x) + f(x) \quad \text{for } x \in \Omega_\varepsilon,$$

$$\operatorname{div}(-a_\varepsilon(x)\nabla v_\varepsilon(x)) = u_\varepsilon(x) - v_\varepsilon(x) + g(x) \quad \text{for } x \in \Omega_\varepsilon,$$

together with the boundary conditions

$$u_\varepsilon = 0 \quad \text{at } \partial\Omega_\varepsilon,$$

$$v_\varepsilon = 0 \quad \text{at } \partial\Omega_\varepsilon.$$

(i) Specify the conditions needed for the matrix a_ε and for the functions f, g so that the Lax–Milgram lemma ensures the existence and uniqueness of a weak solution $(u_\varepsilon, v_\varepsilon)$ to (P_ε). Justify your statement.

(ii) Prove that $(u_\varepsilon, v_\varepsilon)$ is uniformly bounded in $H_0^1(\Omega_\varepsilon) \times H_0^1(\Omega_\varepsilon)$.

(iii) Show that $(u_\varepsilon, v_\varepsilon) \in H_0^1(\Omega_\varepsilon) \times H_0^1(\Omega_\varepsilon)$ can be extended to $(\tilde{u}_\varepsilon, \tilde{v}_\varepsilon)$, which is actually bounded in $H^1(\Omega) \times H^1(\Omega)$.

(iv) Pass in (P_ε) via two-scale convergence to the limit $\varepsilon \to 0$, and determine the weak form of the two-scale limit problem, say (P_0), with solution (u_0, v_0).

(v) Prove the existence and uniqueness of the weak solution to (P_0).

(vi) Get the strong formulation of (P_0). Eliminate u_1, v_1 and show the result.

(vii) Argue in two different ways why $u_0, v_0 \in L^1(\Omega)$, i.e. the system contains "finite mass".

(viii) Take now $\Gamma_\varepsilon^N \neq \emptyset$, and consider the same (P_ε) now with homogeneous Neumann boundary conditions for both observables $u_\varepsilon, v_\varepsilon$ (on part of Γ_ε). Take as solution space

$$V_\varepsilon := \{\varphi \in H^1(\Omega_\varepsilon) : \varphi = 0 \text{ at } \partial\Omega \cup \Gamma_\varepsilon\}.$$

What changes now in the homogenization procedure for this modified problem (P_ε)? Take instead of a_ε a coefficient b_ε (still fulfilling the hypotheses of Lax-Milgram Lemma) so that $b_\varepsilon \neq a_\varepsilon$. What changes now in the homogenization procedure for this modified problem (P_ε)? Justify your answers to both questions.

Exercise 4.8.15. Show how does ε enter the following model equations which describe a standard diffusion scenario in a reactive porous media: Two chemical species diffuse through a periodically structured porous media and react chemically at the surface of the pores producing a third chemical compound that is immobile at those interfaces. Start from the same perforated geometry described in Section 4.2 (the "perforations" are now "pores"), with $Z = [0, \ell]^3$, Y_0 a ball centered in Z, $\Gamma = \partial Y_0$, and let $\Omega = [0, \omega]^3$, where $\ell, \omega \in (0, \infty)$ with $\ell \ll \omega$. Introduce the volumetric mass concentrations u, v, corresponding to the diffusing chemical species U, V, as well as the surface mass concentration z corresponding to the deposited chemical Z. Using the $\{M, L, T\}$ system of units, we have $[U] = [V] = ML^{-3}$ and $[Z] = ML^{-2}$. Let u, v, z be the unknowns of the system having variables x, t, with $[x] = L$ and $[t] = T$. We make use of a number of parameters, $A(\cdot), B(\cdot), \alpha, \eta, u_I, v_I$ and z_I, while the model equations formulated at the pore level $Y = Z - \bar{Y}_0$ read

$$\partial_t u + \nabla \cdot \big(-A\nabla u \big) = 0 \quad \text{in } Y, \tag{4.61}$$

$$\partial_t v + \nabla \cdot \big(-B\nabla v \big) = 0 \quad \text{in } Y, \tag{4.62}$$

$$-A(x)\nabla u \cdot n(x) = -\eta(u, v) \quad \text{at } \Gamma, \tag{4.63}$$

$$-B(x)\nabla v \cdot n(x) = -\eta(u, v) \quad \text{at } \Gamma, \tag{4.64}$$

$$\frac{\mathrm{d}z}{\mathrm{d}t} = \alpha\eta(u, v) \quad \text{at } \Gamma, \tag{4.65}$$

$$z(0, x) = z_I(x), x \in \Gamma; \quad u(0, x) = u_I(x); \quad v(0, x) = v_I(x), x \in \mathrm{clos}(Y). \tag{4.66}$$

Assume that we have homogeneous Neumann boundary conditions. To fix ideas, for any fixed $x \in \Gamma$, take $\eta(x, \cdot)$: $\mathbb{R}^2 \to \mathbb{R}$, given by

$$\eta(x, r, s) = k(x)(r^+)^p(s^+)^q,$$

with $p, q \in \mathbb{R}$, while $r^+ := \max\{r, 0\}$.

To perform this scaling exercise, introduce a characteristic length scale x_{ref}, a characteristic time scale t_{ref}, and reference parameters such as $U(T, X) = \frac{u_{\text{ref}}(x,t)}{u_{\text{ref}}}$, where $X = \frac{x}{x_{\text{ref}}}$ and $T = \frac{t}{t_{\text{ref}}}$. Mind that we expect $\varepsilon = \frac{\ell}{\omega}$, but can you also find other interpretations of this small parameter ε? Depending on your definition of ε, formulate the dimensionless microscopic problem scaled in terms of the parameter ε.

Exercise 4.8.16. Make the following thought experiment. Imagine that a large population of O_2 molecules diffuse through the heterogeneous medium, as depicted in Figure 4.5. Consider that this choice of geometry can be approximated sufficiently well by a periodically perforated geometry. Write for this new domain the mass balance of O_2 (use either the mass or volume concentration of O_2), knowing additionally that:

(i) the blood vessels are variable sources for O_2 such that (i) O_2 crosses the boundary of the blood vessels and enters with no resistance the perforated domain, and (ii) the bigger the diameter of the blood vessel, the weaker the O_2 pump;
(ii) O_2 is consumed inside the perforated domain via a volumetric production term of Michaelis–Menten type;

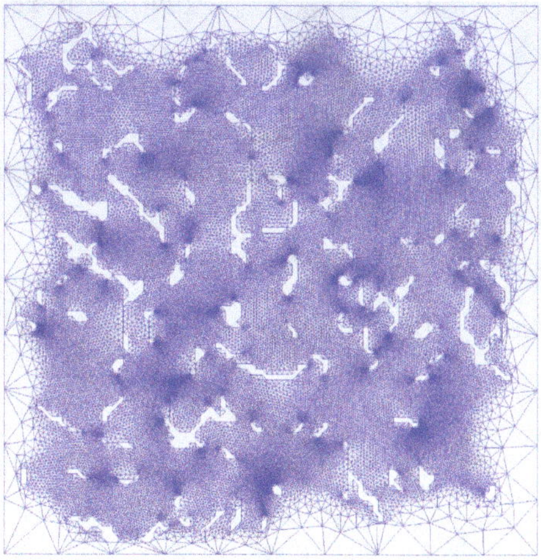

Figure 4.5. Cross-section view of the geometry of a cancer region; see Korhonen *et al.* (2020) for more details. It is a slice of pancreatic islet cell carcinoma in transgenic RIP-Tag mice. The white spots are the blood vessels, while the reminder is the domain where the O_2 is supposed to travel.

(iii) the perforated domain has a uniform initial distribution of oxygen.

Consider that this scenario describes the microscopic problem of O_2 diffusion and reaction inside the perforated domain. Non-dimensionalize your model equations, introduce the small homogenization parameter ε, and finally, formulate the model equations of the microscopic problem, say (P_ε).

Exercise 4.8.17. Let $\Omega \subset \mathbb{R}^d$ be a non-empty domain with Lipschitz boundary. Taking over the microscopic geometry and the spirit of its notation from Chechkin (2021), we introduce a hypersurface γ_ε placed at distance ε from $\partial\Omega$. Fix on this hypersurface the centers of M_ε disjoint balls, say B_j^ε with radius proportional to ε; e.g. take $\rho_j\varepsilon$ with $\rho_j \in (0, 1/2)$ for a set of indices $J := \{1, \ldots, M_\varepsilon\}$. Denote $H_\varepsilon := \cup_{j\in J} B_j^\varepsilon$ and $\Gamma_\varepsilon := \partial H_\varepsilon$. Here, we have in mind a sequence $(M_\varepsilon) \subset \mathbb{Z}$, with $\lim_{\varepsilon\to 0} M_\varepsilon = +\infty$. We set the perforated domain to be $\Omega_\varepsilon := \Omega\backslash\bar{H}_\varepsilon$, and take $\Gamma_\varepsilon = \partial H_\varepsilon$. We do not detail here further considerations on the perforated domain, but some additional geometric constraints are needed for the mathematical arguments to make sense, such as the inequalities (7) or (8) reported by Chechkin (2021).

We are interested in the weak solution $(u_\varepsilon, v_\varepsilon) \in H_0^1(\Omega_\varepsilon)\times H_0^1(\Omega_\varepsilon)$ to the following linear and coupled elliptic system, referred here as problem (P_ε):

$$\begin{aligned} \operatorname{div}(-a_\varepsilon \nabla u_\varepsilon) &= v_\varepsilon - u_\varepsilon + \operatorname{div} \mathrm{b}_1 \quad \text{in } \Omega_\varepsilon, \\ \operatorname{div}(-a_\varepsilon \nabla v_\varepsilon) &= u_\varepsilon - v_\varepsilon + \operatorname{div} \mathrm{b}_2 \quad \text{in } \Omega_\varepsilon, \\ u_\varepsilon &= v_\varepsilon = 0 \quad \text{at } \partial\Omega_\varepsilon. \end{aligned}$$

The mathematical structure of the coupling appears in the context of concentrations diffusion involved in gaseous-liquid chemical reactions taking place in porous media.

We assume the following restrictions on data:

(H1) There exist $\alpha, \beta \in (0, \infty)$ independent of ε (with $\alpha < \beta$) such that condition

$$\alpha||\zeta||^2_{\mathbb{R}^d} \leq \sum_{i,j=1}^{d} A_{ij}\left(\frac{x}{\varepsilon}\right)\zeta_i\zeta_j \leq \beta||\zeta||^2_{\mathbb{R}^d} \tag{4.67}$$

holds for almost all $x \in \Omega_\varepsilon$ and for all $\zeta := (\zeta_1, \ldots, \zeta_d) \in \mathbb{R}^d$, where $a_\varepsilon(x) := A\left(\frac{x}{\varepsilon}\right) \in \mathbb{R}^{d\times d}$;

(H2) $b_1, b_2 \in W^{1,2+\delta_0}(\Omega)$, with $\delta_0 > 0$ so that $B_1 \neq B_2$;

(H3) a_ε is uniformly bounded in $L^\infty(\Omega_\varepsilon)$.

Given (H1)–(H3), handle the following tasks:

(i) Prove the weak solvability of problem (P_ε). Show that the desired weak solution is also unique and stable with respect to all model parameters.

(iii) Prove that there exist two constants $\delta > \delta_0$ and $c > 0$ (both independent of ε) such that the following estimate holds:

$$\int_{\Omega_\varepsilon} |\nabla u_\varepsilon|^{2+\delta} dx + \int_{\Omega_\varepsilon} |\nabla v_\varepsilon|^{2+\delta} dx \leq c. \tag{4.68}$$

(ii) Pass with $\varepsilon \to 0$ and identify the model equations in terms of the unknowns u_0 and v_0, corresponding to the limit problem (P_0). Ensure the well-posedness of problem (P_0) in the weak sense.

(iv) Prove a corrector estimate of the type: There exists a constant $c^* > 0$ (independent of ε) such that the following estimate holds:

$$||u_\varepsilon - \hat{u}_0||_{H_0^1(\Omega)} + ||v_\varepsilon - \hat{v}_0||_{H_0^1(\Omega)} \leq c^* \varepsilon^\theta, \tag{4.69}$$

where $\hat{\phi}$ denotes the zero Sobolev extension of the associated function ϕ. Give a direct proof without using (4.68), and then redo the proof this time involving (4.68). Point out (as explicitly as you can) the pair (c^*, θ) for these situations.

(v) Look at Theorem 4.1 from Ming and Song (2024), and assume an estimate of type (4.69) to be available in their setting. How would then the exponent θ affect the right-hand side of the convergence error estimate (4.1) and (4.2) in *loc. cit.*? Is it improving the convergence rate?

Exercise 4.8.18. Let $S = (0, T)$ be a time interval and $\emptyset \neq \Omega_\varepsilon \subset \mathbb{R}^d$ be a perforated domain as in the previous exercises. We consider the following parabolic problem modeling a diffusion-chemical reaction process in the homogeneous domain $\Omega_T^\varepsilon := S \times \Omega_\varepsilon$ given by

$$b_\varepsilon(x)\partial_t u_\varepsilon(t,x) + \mathrm{div}\big(-a_\varepsilon(x)\nabla u_\varepsilon(t,x)\big) = c_\varepsilon(x) f(t,x) \quad \text{for } (t,x) \in \Omega_T^\varepsilon$$

$$u_\varepsilon(t,x) = 0 \quad \text{for } (t,x) \in S \times \partial\Omega_\varepsilon$$

$$u_\varepsilon(0,x) = d(x) \quad \text{for } x \in \bar{\Omega}_\varepsilon,$$

where $a_\varepsilon(x) = A(\frac{x}{\varepsilon})$ satisfies the usual coercivity assumption, $b(\cdot)$ is a bounded positive Y-periodic function such that

$$0 < b^- \leq b(y) \leq b^+ < +\infty$$

for all $y \in Y$, $c \in L^2_\#(Y)$, $d \in L^2(\Omega_\varepsilon)$, and $f \in L^2(\Omega_T^\varepsilon)$. Use the concept of two-scale convergence adapted to time-dependent settings to derive the homogenized version of this problem posed in a perforated domain.

4.9 Solutions

Solution 4.8.8. The solution to this exercise is discussed in detail in Chapter 6 for the case of $\lambda^{d-1}(\Gamma_\varepsilon^D) = 0$. It is worth noting that this type of exercise can be offered as individual course projects.

Solution 4.8.9. Corrector estimates are quantitative inequalities in suitable norms that estimate the quality of the averaging process. An excellent inspiration source for handling this exercise is Section 2.6 in the monograph by Ciorănescu and Saint Jean Paulin (1998).

Solution 4.8.12. Adapt the proof from Ciorănescu and Saint Jean Paulin (1979, Theorem 3) to this setting.

Solution 4.8.13. It is interesting to note first that if $h_\varepsilon = 0$ across $\partial\Omega$, then it results that $u_\varepsilon = 0$ in $\overline{\Omega}_\varepsilon$ is a solution to the proposed system. It only remains to show that in this case, the zero solution is unique. Correspondingly, $\lim_{\varepsilon\to 0} u_\varepsilon(x) = u_0(x) = 0$ for all $x \in \overline{\Omega}$.

The exercise is, of course, more interesting if $h_\varepsilon \neq 0$ across $\partial\Omega$. Assume, for instance, that there exists a set of nonzero measures $\Sigma \subset \partial\Omega$ such that $h_\varepsilon(x) \neq 0$ for all $x \in \Sigma$. This is the case we will discuss next.

To homogenize the Dirichlet boundary condition, we set

$$v_\varepsilon(x) := u_\varepsilon(x) - \varepsilon^3 h_\varepsilon(x).$$

This leads to the following reformulation of the original problem: Find v_ε in the function space[6] V_ε such that the following boundary-value

[6]Observe the link between the functions spaces V_ε and $\mathcal{V}_\varepsilon$, that is, $\mathcal{V}_\varepsilon := \varepsilon^3 h_\varepsilon + V_\varepsilon$.

problem is satisfied:

$$\begin{aligned}\operatorname{div}\big(-a_\varepsilon(x)\nabla v_\varepsilon(x)\big) + \alpha v_\varepsilon(x) = f_\varepsilon(x) \quad &\text{for } x \in \Omega_\varepsilon,\\ -a_\varepsilon(x)\nabla v_\varepsilon(x)\cdot n_\varepsilon(x) = \varepsilon v_\varepsilon(x) + \varepsilon g_\varepsilon(x) \quad &\text{for } x \in \Gamma_\varepsilon,\\ v_\varepsilon(x) = 0 \quad &\text{for } x \in \partial\Omega,\end{aligned}$$

where

$$\begin{aligned} f_\varepsilon(x) &:= -\alpha\varepsilon^3 h_\varepsilon(x) + \varepsilon^3 \operatorname{div}(a_\varepsilon(x)\nabla h_\varepsilon(x)),\\ g_\varepsilon(x) &:= \varepsilon^3 h_\varepsilon(x) + \varepsilon^2 a_\varepsilon(x)\nabla h_\varepsilon(x)\cdot n_\varepsilon(x).\end{aligned}$$

Before starting to answer the indicated questions, it makes sense to explore a bit what options we have for concrete choices of periodically oscillating functions h_ε such that, on the one hand, we avoid non-homogenizable situations and, on the other hand, the resulting homogenized problem admits non-zero solutions.

Solution 4.8.14. The quickest way to solve this exercise is to use what we have seen in the one-dimensional example discussed in Section 2.1. Alternatively, a direct approach would work as well.

Solution 4.8.15. In the $\{M, L, T\}$ system of units, we have $[x] = L$, $[t] = T$, $[u] = [v] = ML^{-3}$, and $[z] = ML^{-2}$. Since $[\partial_t] = T^{-1}$ and $[\partial_x] = L^{-1}$, we find from dimensional analysis that $[A] = [B] = L^2T^{-1}$, $[\alpha] = [p] = [q] = 1$, and $[k] = M^{1-p-q}L^{3p+3q-2}T^{-1}$. The meaning of this bracket $[\cdot]$ is explained carefully by Pawlowski (1971). Here, we only mention that the notation $[r]$ points out the units of the quantity r. To nondimensionalize the system (4.61)–(4.66), we introduce a couple of scaling factors linking dimensional with dimensionless quantities:

$$t = t_{\text{ref}}\tilde{t}, \tag{4.70}$$

$$x = x_{\text{ref}}\tilde{x}, \tag{4.71}$$

$$u(x,t) = u_{\text{ref}}\tilde{u}(\tilde{x},\tilde{t}), \tag{4.72}$$

$$v(x,t) = v_{\text{ref}}\tilde{v}(\tilde{x},\tilde{t}), \tag{4.73}$$

$$z(x,t) = z_{\text{ref}}\tilde{z}(\tilde{x},\tilde{t}), \tag{4.74}$$

$$A(x) = A_{\text{ref}}\tilde{A}(\tilde{x}), \tag{4.75}$$

$$B(x) = B_{\text{ref}}\tilde{B}(\tilde{x}), \tag{4.76}$$

$$k(x) = k_{\text{ref}}\tilde{k}(\tilde{x}), \tag{4.77}$$

$$\tilde{Y} = \frac{Y}{x_{\text{ref}}}, \tag{4.78}$$

$$\tilde{\Gamma} = \frac{\Gamma}{x_{\text{ref}}}, \tag{4.79}$$

$$\tilde{\tilde{Y}} = \frac{\tilde{Y}}{x_{\text{ref}}}, \tag{4.80}$$

$$n(x) = \tilde{n}(\tilde{x}). \tag{4.81}$$

Inserting (4.70)–(4.81) into (4.61)–(4.66) yields

$$\partial_{\tilde{t}}\tilde{u} + \frac{A_{\text{ref}}t_{\text{ref}}}{x_{\text{ref}}^2}\tilde{\nabla}\cdot\left(-\tilde{A}(\tilde{x})\tilde{\nabla}\tilde{u}\right) = 0 \quad \text{in } \tilde{Y}, \tag{4.82}$$

$$\partial_{\tilde{t}}\tilde{v} + \frac{B_{\text{ref}}t_{\text{ref}}}{x_{\text{ref}}^2}\tilde{\nabla}\cdot\left(-\tilde{B}(\tilde{x})\tilde{\nabla}\tilde{v}\right) = 0 \quad \text{in } \tilde{Y}, \tag{4.83}$$

$$-\tilde{A}(\tilde{x})\tilde{\nabla}\tilde{u}\cdot\tilde{n}(\tilde{x}) = -\frac{x_{\text{ref}}k_{\text{ref}}u_{\text{ref}}^p v_{\text{ref}}^q}{A_{\text{ref}}u_{\text{ref}}}\eta(\tilde{u},\tilde{v}) \quad \text{at } \tilde{\Gamma}, \tag{4.84}$$

$$-\tilde{B}(\tilde{x})\tilde{\nabla}\tilde{v}\cdot\tilde{n}(\tilde{x}) = -\frac{x_{\text{ref}}k_{\text{ref}}u_{\text{ref}}^p v_{\text{ref}}^q}{B_{\text{ref}}v_{\text{ref}}}\eta(\tilde{u},\tilde{v}) \quad \text{at } \tilde{\Gamma} \tag{4.85}$$

$$\frac{\mathrm{d}\tilde{z}}{\mathrm{d}\tilde{t}} = \alpha\frac{t_{\text{ref}}k_{\text{ref}}u_{\text{ref}}^p v_{\text{ref}}^q}{z_{\text{ref}}}\eta(\tilde{u},\tilde{v}) \quad \text{at } \tilde{\Gamma}, \tag{4.86}$$

$$\tilde{z}(\tilde{t}=0,\tilde{x}) = \frac{z_I(x_{\text{ref}}\tilde{x})}{z_{\text{ref}}}, \quad \tilde{x}\in\tilde{\Gamma}; \tag{4.87}$$

$$\tilde{u}(\tilde{t}=0,\tilde{x}) = \frac{u_I(x_{\text{ref}}\tilde{x})}{u_{\text{ref}}}; \quad \tilde{v}(\tilde{t}=0,\tilde{x}) = \frac{v_I(x_{\text{ref}}\tilde{x})}{v_{\text{ref}}} \quad \tilde{x}\in\text{clos}(\tilde{Y}). \tag{4.88}$$

To proceed further, we choose as the characteristic time scale $t_{\text{ref}} = \frac{x_{\text{ref}}^2}{A_{\text{ref}}}$, i.e. we set the reference time as the *diffusion time* of the chemical species U. The evolution system becomes

$$\partial_{\tilde{t}}\tilde{u} + \tilde{\nabla}\cdot\left(-\tilde{A}(\tilde{x})\tilde{\nabla}\tilde{u}\right) = 0 \quad \text{in } \tilde{Y}, \tag{4.89}$$

$$\partial_{\tilde{t}}\tilde{v} + \frac{B_{\text{ref}}}{A_{\text{ref}}}\tilde{\nabla}\cdot\left(-\tilde{B}(\tilde{x})\tilde{\nabla}\tilde{v}\right) = 0 \quad \text{in } \tilde{Y}, \tag{4.90}$$

$$-\tilde{A}(\tilde{x})\tilde{\nabla}\tilde{u}\cdot\tilde{n}(\tilde{x}) = -\frac{x_{\text{ref}}k_{\text{ref}}u^p_{\text{ref}}v^q_{\text{ref}}}{A_{\text{ref}}u_{\text{ref}}}\eta(\tilde{u},\tilde{v}) \quad \text{at } \tilde{\Gamma}, \tag{4.91}$$

$$-\tilde{B}(\tilde{x})\tilde{\nabla}\tilde{v}\cdot\tilde{n}(\tilde{x}) = -\frac{x_{\text{ref}}k_{\text{ref}}u^p_{\text{ref}}v^q_{\text{ref}}}{B_{\text{ref}}v_{\text{ref}}}\eta(\tilde{u},\tilde{v}) \quad \text{at } \tilde{\Gamma} \tag{4.92}$$

$$\frac{\mathrm{d}\tilde{z}}{\mathrm{d}\tilde{t}} = \alpha\frac{x^2_{\text{ref}}k_{\text{ref}}u^p_{\text{ref}}v^q_{\text{ref}}}{A_{\text{ref}}z_{\text{ref}}}\eta(\tilde{u},\tilde{v}) \quad \text{at } \tilde{\Gamma}, \tag{4.93}$$

$$\tilde{z}(\tilde{t}=0,\tilde{x}) = \frac{z_I(x_{\text{ref}}\tilde{x})}{z_{\text{ref}}}, \quad \tilde{x}\in\tilde{\Gamma}; \tag{4.94}$$

$$\tilde{u}(\tilde{t}=0,\tilde{x}) = \frac{u_I(x_{\text{ref}}\tilde{x})}{u_{\text{ref}}}; \quad \tilde{v}(\tilde{t}=0,\tilde{x}) = \frac{v_I(x_{\text{ref}}\tilde{x})}{v_{\text{ref}}} \quad x\in\text{clos}(\tilde{Y}). \tag{4.95}$$

To introduce the small parameter in the dimensionless model equations, we set $\varepsilon = \frac{\ell}{\omega}$ and impose the constraint of the typical size of the pore (or of the microstructure), i.e.

$$\frac{1}{\varepsilon}\text{diam}(\tilde{Y}) = \mathcal{O}(1). \tag{4.96}$$

Introducing the ε-notation, accounting for the space periodicity of the space-dependent parameters and model functions, (4.89)–(4.95), yields

$$\partial_{\tilde{t}}\tilde{u}_\varepsilon + \tilde{\nabla}\cdot\left(-\tilde{A}\left(\frac{\tilde{x}}{\varepsilon}\right)\tilde{\nabla}\tilde{u}_\varepsilon\right) = 0 \quad \text{in } \tilde{\Omega}_\varepsilon, \tag{4.97}$$

$$\partial_{\tilde{t}}\tilde{v}_\varepsilon + \frac{B_{\text{ref}}}{A_{\text{ref}}}\tilde{\nabla}\cdot\left(-\tilde{B}\left(\frac{\tilde{x}}{\varepsilon}\right)\tilde{\nabla}\tilde{v}_\varepsilon\right) = 0 \quad \text{in } \tilde{\Omega}_\varepsilon, \tag{4.98}$$

$$-\tilde{A}\left(\frac{\tilde{x}}{\varepsilon}\right)\tilde{\nabla}\tilde{u}_\varepsilon\cdot\tilde{n}_\varepsilon = -\frac{x_{\text{ref}}k_{\text{ref}}u^p_{\text{ref}}v^q_{\text{ref}}}{A_{\text{ref}}u_{\text{ref}}}\eta(\tilde{u}_\varepsilon,\tilde{v}_\varepsilon) \quad \text{at } \tilde{\Gamma}_\varepsilon, \tag{4.99}$$

$$-\tilde{B}\left(\frac{\tilde{x}}{\varepsilon}\right)\tilde{\nabla}\tilde{v}_\varepsilon\cdot\tilde{n}_\varepsilon = -\frac{x_{\text{ref}}k_{\text{ref}}u^p_{\text{ref}}v^q_{\text{ref}}}{B_{\text{ref}}v_{\text{ref}}}\eta(\tilde{u}_\varepsilon,\tilde{v}_\varepsilon) \quad \text{at } \tilde{\Gamma}_\varepsilon, \tag{4.100}$$

$$\frac{\mathrm{d}\tilde{z}_\varepsilon}{\mathrm{d}\tilde{t}} = \alpha\frac{x^2_{\text{ref}}k_{\text{ref}}u^p_{\text{ref}}v^q_{\text{ref}}}{A_{\text{ref}}z_{\text{ref}}}\eta(\tilde{u}_\varepsilon,\tilde{v}_\varepsilon) \quad \text{at } \tilde{\Gamma}_\varepsilon, \tag{4.101}$$

$$\tilde{z}_\varepsilon(\tilde{t}=0) = \frac{1}{z_{\text{ref}}}z_I\left(\frac{\tilde{x}}{\varepsilon}\right) \quad \tilde{x}\in\tilde{\Gamma}_\varepsilon \tag{4.102}$$

$$\tilde{u}_\varepsilon(\tilde{t}=0) = \frac{1}{u_{\text{ref}}}u_I\left(\frac{\tilde{x}}{\varepsilon}\right); \quad \tilde{v}_\varepsilon(\tilde{t}=0) = \frac{1}{v_{\text{ref}}}v_I\left(\frac{\tilde{x}}{\varepsilon}\right) \quad x\in\text{clos}(\tilde{\Omega}_\varepsilon). \tag{4.103}$$

Interestingly, we obtain four different dimensionless numbers that can in principle scale very differently in terms of ε. These dimensionless numbers are

$$\frac{B_{\text{ref}}}{A_{\text{ref}}} = \varepsilon^{\gamma_1}, \tag{4.104}$$

$$\frac{x_{\text{ref}} k_{\text{ref}} u_{\text{ref}}^p v_{\text{ref}}^q}{A_{\text{ref}} u_{\text{ref}}} = \varepsilon^{\gamma_2}, \tag{4.105}$$

$$\frac{x_{\text{ref}} k_{\text{ref}} u_{\text{ref}}^p v_{\text{ref}}^q}{B_{\text{ref}} v_{\text{ref}}} = \varepsilon^{\gamma_3}, \tag{4.106}$$

$$\frac{x_{\text{ref}}^2 k_{\text{ref}} u_{\text{ref}}^p v_{\text{ref}}^q}{A_{\text{ref}} z_{\text{ref}}} = \varepsilon^{\gamma_4}, \tag{4.107}$$

with $\gamma_1, \gamma_2, \gamma_3, \gamma_4 \in \mathbb{R}$. In chemical engineering, they are usually referred to as Damköhler numbers. The four different dimensionless numbers capture the same physics of the situation from four different perspectives.

As for now, the choices of γ_i for $i = \{1, \ldots, 4\}$ are arbitrary. The variations in the choices are limited by the physics of the situation. Values for these exponents can eventually be pinned down by laboratory experiments. However, we ask ourselves how this identification of exponents should take place *de facto*? This is a simple natural question, which is not easy to address in a satisfactory manner. Two related concrete questions are the following:

- Should laboratory experiments reveal *microscopic* or *macroscopic* information?
- What type of information is observable at which time and length scale?

These types of questions directly connect multiscale techniques to topics in non-standard inverse problems.[7] A typical choice of good exponents is $(\gamma_1, \gamma_2, \gamma_3, \gamma_4) = (0, 1, 1, 0)$. What other options of exponents do you think

[7] We do not touch the inverse problems research direction within the framework of this textbook, but it is certainly a very promising one, capable of shedding light on concrete practical questions, employing, at the same time, a nice combination of functional analysis-type reasoning, asymptotic techniques, and computational work. Multiscale inverse problems can be formulated for larger projects at the MSc or PhD thesis level.

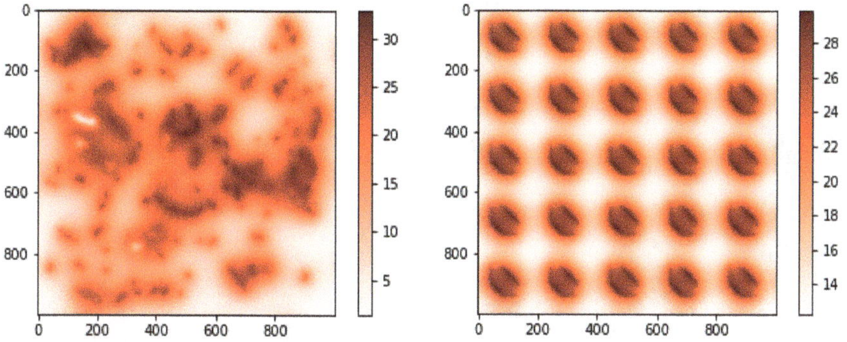

Figure 4.6. Cross-section view of the geometry of the cancer region shown in Figure 4.5 – a periodic approximation.

are possible to handle mathematically from the perspective of two-scale convergence if one assumes beforehand the case of a linear production by reaction $\eta(\cdot)$?

Solution 4.8.16. There are a couple of tasks that need to be handled. The first one is to approximate the geometry described in Figure 4.5 with a periodic (or locally periodic) representation, as in Figure 4.6, for example. This gives an idea about the choice of ε, as well as about typical realistic sizes. Then, one has to translate the imposed physical description in terms of balance equations.

Solution 4.8.17. We only give a hint about the solution. Standard application of the Lax–Milgram lemma leads to (i). To prove (ii), we denote $w_\varepsilon := u_\varepsilon + v_\varepsilon$ and $p_\varepsilon := u_\varepsilon - v_\varepsilon$ and apply for the resulting equations in w_ε and p_ε the structure of the Bojarski–Meyers estimate (cf. Theorem 1 in Chechkin (2021)). Scaling and weak convergence arguments clarify (iii). The proof ideas from Section 4 in Chechkin (2021) apply to the partial differential equations formulated for w_ε and p_ε and lead to (iv). This is a corrector estimate of the homogenization process. One easily sees the use of (4.69) in the structure of the desired inequality in (v). The Bojarski–Meyers inequality improves slightly the quality of the convergence rate.

Solution 4.8.18. This type of exercise aligns with the techniques developed for evolution problems, similar to those discussed in the following chapter. A well-written methodological paper in this direction is that by Peter and Böhm (2008).

Chapter 5

Interplaying Limits and Preservation of Cultural Heritage: Dimension Reduction, Periodic Homogenization, and Fast-Reaction Asymptotics

In this chapter, we explore a scenario relevant to the preservation of cultural heritage. Beyond the novelty of the application field, the key new element introduced is the adaptation of the two-scale convergence concept to time-dependent settings. Additionally, we incorporate concepts such as dimension reduction and fast-reaction asymptotics in a reaction-diffusion-adsorption model that applies to concrete-based historical buildings.

After a brief introduction to the problem setting of concrete carbonation, we use mathematical analysis tools to examine the corresponding reaction-diffusion process in a two-phase medium with a microscopic length scale ε. The diffusion coefficients in the two phases are assumed to be highly different ($d_1/D = \varepsilon^2$), while the reaction constant k is large. Such scaling facilitates the formation of sharp fronts that can quickly penetrate the material. First, we take the homogenization limit $\varepsilon \to 0$ and derive a two-scale model. Afterward, we pass to the fast-reaction limit $k \to \infty$ and obtain a two-scale reaction-diffusion system with a moving boundary traveling within the underlying microstructure.

The key questions we ask here are as follows:

(i) What is the microscopic origin of the formation of sharp moving fronts?
(ii) How does dimension reduction interplay with other asymptotics such as homogenization and/or fast-reaction limits?

Most of the material presented in this chapter was originally published in the work of Meier and Muntean (2010). We revisit it here, expanding its original scope. In particular, as highlighted by Muntean (2023), we aim to emphasize that such mathematical investigations can be applied not only to materials science (where the porous structure of underlying fabrics can be easily exploited) but also in the context of cultural heritage preservation.

For more related details about this application, we refer the reader to Bonetti *et al.* (2021) for a relatively recent collection of attempts from the side of the mathematical community to contribute to non-invasive investigations of cultural heritage (broadly seen).

5.1 A Multiscale Perspective into the Preservation of Cultural Heritage: Understanding Chemical Reactions Occurring in Heterogeneous Multi-Phase Porous Media

The perspective we adopt here was introduced by Muntean (2023). We believe that mathematical investigations can provide valuable insights, not only in materials science (where the porous structure of underlying fabrics must be analyzed in detail) but also in cultural heritage preservation, where microstructures are often challenging to analyze non-destructively. However, the reader may wonder what the carbonation reaction has to do with cultural heritage preservation and why a multiscale approach is needed. The first part of the question is addressed by Courard *et al.* (2021), where the authors state, "Among building structures, concrete monuments, churches, and houses are a significant part of our cultural heritage, representing an architectural period that was both promising and enthusiastic for the social development of humanity". As for the second part of the question, we highlight the fact that cementitious materials (including concrete) are typically heterogeneous and multiphase, containing an intricate network of interlinked pathways that facilitate both the transport and storage of matter. A crucial point is that the carbonation reaction does not destroy the porous fabric itself; it only alters the local porosity and, consequently, lowers the local pH. As a result, the material becomes more vulnerable to ion attacks from the environment (e.g. sulfates, chlorides), negatively affecting both its cultural value and long-term durability. To fix ideas, we refer the reader to our work (Jävergård *et al.*, 2025), where we propose a hybrid (continuous-stochastic) model describing the degradation of marble monuments due to the ingress of sulfate ions.

To capture multiscale effects, we begin by assuming that culturally valuable materials possess a relatively regular internal structure, allowing them to be effectively approximated using periodic or locally periodic perforated media. The next step is to consider a semi-linear parabolic system posed on such perforated media to model a large class of gas–solid reactions, including the carbonation reaction. The physical scenario we have in mind is as follows: A gaseous species A penetrates a non-saturated porous medium via the air phase of its pore space, dissolves in the pore water, and reacts rapidly with a species B. For simplicity, we assume that species B is already present in the pore water at the start. In general, however, species B will become available through a dissolution mechanism or any other depletion process that transfers B from the solid matrix into the pore solution.

It is known that the homogenization limit of the corresponding pore-scale model has a two-scale structure, provided that diffusion of A in the pore water is sufficiently slow (Hornung and Jäger, 1991; Hornung *et al.*, 1994). On the other hand, regarding the fast-reaction asymptotics, we are interested in the singular-limit analysis $k \to \infty$, where k is the reaction constant. In our work (Meier and Muntean, 2008), we have shown by formal asymptotics that it does not matter in which order we take limits, or, in other words, whether we first pass to the fast-reaction limit $k \to \infty$ and then take the homogenization limit $\varepsilon \to 0$ or *vice versa*. If no diffusion of B is allowed, the expected result is a two-scale problem with a family of one-phase Stefan problems on the micro scale.

In the remainder of this chapter, we plan to prove rigorously the singular limit analyses $\varepsilon \to 0$ and then $k \to \infty$ (the order is now fixed). To simplify the mathematical analysis, we allow for the diffusion of B in the pore water and replace the instantaneous (Dirichlet) exchange condition between the two phases by a regularizing Robin condition.

We perform our asymptotic analysis in two stages:

(i) Firstly, we use a standard homogenization-based approach (very much inspired by the works of Hornung and Jäger (1991), Hornung *et al.* (1994), and Meier (2008)) to prove the convergence as $\varepsilon \to 0$ of the starting micro PDE system toward a two-scale reaction-diffusion system.
(ii) Secondly, we adapt to the setting of two-scale parabolic equations some of the fast-reaction asymptotics techniques developed by Fasano *et al.* (1990), Dancer *et al.* (1999), Crooks *et al.* (2004), Hilhorst *et al.* (2012), and Seidman (2009) and then prove the limiting behavior $k \to \infty$.

We refer the reader to Nishiura (2002), Meier *et al.* (2007), and van Noorden and Pop (2008) for a collection of fast-reaction–(slow)-diffusion settings playing an important role in pattern formation and corrosion of porous materials and to Friedman and Tzavaras (1987) and Murad and Cushman (1996) for more conceptually related scenarios arising in the modeling of catalytic reactors and deformation in hydrophilic swelling porous media.

5.2 The Microscopic Problem and a Dimension Reduction Argument

Let $L > 0$ and $R \in (0, 1)$ be given lengths, and let $\varepsilon > 0$ be a small number. We consider a layered medium $G := (0, L)^2$ in two space dimensions divided into periodic cells εZ, which are ε-scaled versions of the unit cell $Z := (0, \frac{1}{2}) \times (-1, 1)$, as depicted in Figure 5.1. Let $Z_2 := (0, \frac{1}{2}) \times (-R, R)$

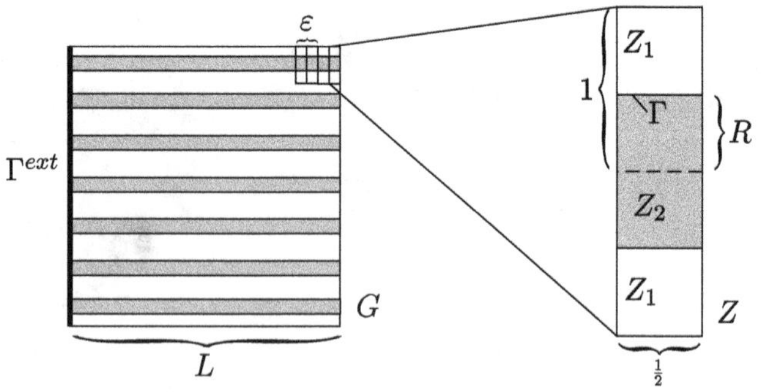

Figure 5.1. Geometry of the model (P_k^ε). The right-hand side figure is scaled by ε^{-1}.

be the material where diffusion is slow and $Z_1 := Z\backslash\overline{Z}_2$ be the remaining part. Then, G is separated into subdomains:

$$G_i^\varepsilon := G \cap \bigcup_{(2k_1, \frac{k_2}{2})\in\mathbb{Z}^2} \varepsilon(k + Z_i), \qquad i \in \{1, 2\}.$$

The interface between the two phases is denoted by $\Gamma^\varepsilon := \partial G_1^\varepsilon \cap \partial G_2^\varepsilon$. Its unit normal pointing toward G_1^ε is $\nu := (0, \pm 1)^T$. Moreover, we specify two different parts of the exterior boundary ∂G via $\Gamma^{ext} := \{0\} \times (0, L)$ and $\Gamma^N := \partial G\backslash\Gamma^{ext}$.

Let D, d_1, d_2, k, b, and α be positive constants. We consider the following reaction-diffusion problem posed at the level of the microscopic scale:

Find $(U^\varepsilon, u^\varepsilon, v^\varepsilon)$ satisfying

$$\partial_t U^\varepsilon - D\Delta U^\varepsilon = 0 \quad x \in G_1^\varepsilon,\ t > 0, \tag{5.1}$$

$$\partial_t u^\varepsilon - \varepsilon^2 d_1 \Delta u^\varepsilon = -k u^\varepsilon v^\varepsilon \quad x \in G_2^\varepsilon,\ t > 0, \tag{5.2}$$

$$\partial_t v^\varepsilon - \varepsilon^2 d_2 \Delta v^\varepsilon = -\alpha k u^\varepsilon v^\varepsilon \quad x \in G_2^\varepsilon,\ t > 0, \tag{5.3}$$

with interior boundary conditions

$$-D\nabla U^\varepsilon \cdot \nu = -\varepsilon^2 d_1 \nabla u^\varepsilon \cdot \nu \tag{5.4}$$

$$= \varepsilon b(u^\varepsilon - U^\varepsilon) \tag{5.5}$$

$$d_2 \nabla v^\varepsilon = 0 \quad x \in \Gamma^\varepsilon, t \geq 0, \tag{5.6}$$

exterior boundary conditions

$$U^\varepsilon = U^{ext}(t) \quad x \in \partial G_1^\varepsilon \cap \Gamma^{ext},\ t \geq 0, \tag{5.7}$$

$$D\nabla U^\varepsilon = 0 \quad x \in \partial G_1^\varepsilon \cap \Gamma^N,\ t \geq 0, \tag{5.8}$$

$$d_1 \nabla u^\varepsilon = d_2 \nabla v^\varepsilon = 0 \quad x \in \partial G_2^\varepsilon \cap \partial G,\ t \geq 0, \tag{5.9}$$

and initial conditions

$$U^\varepsilon(0, x) = U_0(x_1), \tag{5.10}$$

$$u^\varepsilon(0, x) = u_0(x_1, x_2/\varepsilon), \tag{5.11}$$

$$v^\varepsilon(0, x) = v_0(x_1, x_2/\varepsilon), \quad x = (x_1, x_2) \in G. \tag{5.12}$$

The system (5.10)–(5.12) is referred to as (P_k^ε). For typical fast-reaction–slow-transport scenarios, the parameter k is large and defines a high Thiele modulus, which is in fact one of the Damköhler numbers. The case of $b \to \infty$ corresponds (at this point only only formally) to the Dirichlet case $u^\varepsilon|_{y=R} = U^\varepsilon$.

For a given $T > 0$, let $S := (0,T)$ be a bounded time interval. We also denote $\Omega := (0,L)$ and $Y := (0,R)$. We assume the following restrictions on the data and parameters:

$$U^{ext} \in H^1(S), \quad U_0 \in H^1(\Omega), \quad u_0, v_0 \in H^1(\Omega \times Y).$$

Moreover, there is a constant $C_M > 0$ such that

$$0 \leq U^{ext}, U_0, U_0, v_0 \leq C_M \quad \text{a.e.,} \tag{5.13}$$

and the following compatibility conditions are fulfilled:

$$U^{ext}(0) = U_0(0), \qquad u_0(x_1, R) = U_0(x_1),$$
$$\partial_{x_2} u_0(x_1, 0) = \partial_{x_2} v_0(x_1, 0) = \partial_{x_2} v_0(x_1, R) = 0.$$

We can therefore extend $u_0(x_1, x_2)$ and $v_0(x_1, x_2)$ first by symmetry to $x_2 \in (-R, R)$, afterward by U_0 for $x_2 \in (-1, 1)$, and finally periodically to $x_2 \in \mathbb{R}$.

Note that, due to the symmetry of the layered geometry and of the data, it is actually sufficient to consider only one single strip of height ε. However, in order to apply known homogenization results later on, it is more convenient to work with an ε-independent domain G. The symmetry will be taken into account after the homogenization step.

To write a suitable weak formulation for our problem, we introduce the function space

$$V_D^\varepsilon := \{u \in H^1(G_1^\varepsilon) : \; u = 0 \text{ at } \partial G_1^\varepsilon \cap \Gamma^{ext}\}.$$

The weak formulation of Problem (P_k^ε) reads as follows:

Find the functions $u^\varepsilon \in U^{ext} + L^2(S; V_D^\varepsilon)$ and $U^\varepsilon, V^\varepsilon \in L^2(S; H^1(G_2^\varepsilon))$ such that equations (5.10)–(5.12) are satisfied, and it holds that

$$\frac{d}{dt}\int_{G_1^\varepsilon} u^\varepsilon \phi + \int_{G_1^\varepsilon} D\nabla u^\varepsilon \nabla \phi + \int_{\Gamma^\varepsilon} b(U^\varepsilon - u^\varepsilon)\phi = 0, \tag{5.14}$$

$$\frac{d}{dt}\int_{G_2^\varepsilon} u^\varepsilon\Phi + \int_{G_2^\varepsilon} \varepsilon^2 d_1 \nabla u^\varepsilon \nabla\Phi + \int_{\Gamma^\varepsilon} b(u^\varepsilon - U^\varepsilon)\Phi + \int_{G_2^\varepsilon} k u^\varepsilon v^\varepsilon \Phi = 0, \quad (5.15)$$

$$\frac{d}{dt}\int_{G_2^\varepsilon} v^\varepsilon\Psi + \int_{G_2^\varepsilon} \varepsilon^2 d_2 \nabla v^\varepsilon \nabla\Psi + \int_{G_2^\varepsilon} \alpha k u^\varepsilon v^\varepsilon \Psi = 0 \quad (5.16)$$

for all $\phi \in V_D^\varepsilon$ and $\Phi, \Psi \in H^1(G_2^\varepsilon)$.

The following qualitative results hold *a priori* with the upper bounds independent of the choice of ε.

Proposition 5.1. *Any weak solution is nonnegative and essentially bounded via*

$$0 \leq u^\varepsilon, U^\varepsilon, v^\varepsilon \leq C_M \quad a.e.,$$

where C_M is given by (5.13).

Proposition 5.2. *The following energy estimates hold independently of the choice of ε:*

$$\|U^\varepsilon(t)\|^2_{G_1^\varepsilon} + \|u^\varepsilon(t)\|^2_{G_2^\varepsilon} + \int_0^t \|\nabla U^\varepsilon\|^2_{G_1^\varepsilon} + \int_0^t \|\varepsilon\nabla u^\varepsilon\|^2_{G_2^\varepsilon}$$
$$\leq C(\|U_0\|^2_\Omega + \|u_0\|^2_{\Omega\times Y} + \int_0^t (U^{ext})^2), \quad (5.17)$$

$$\|v^\varepsilon(t)\|^2_{G_2^\varepsilon} + \int_0^t \|\varepsilon\nabla v^\varepsilon\|^2_{G_2^\varepsilon} \leq C\|v_0\|^2_{\Omega\times Y}. \quad (5.18)$$

Proof. Test (5.14) with $\phi := u^\varepsilon - U^{ext}$ and (5.15) with $\Phi := u^\varepsilon$ and adding the result yields (5.17). The second estimate (5.18) is obtained by testing (5.15) with $\Psi := v^\varepsilon$. □

Proposition 5.3. *The following improved estimate holds independently of ε:*

$$\|\partial_t U^\varepsilon(t)\|^2_{G_1^\varepsilon} + \|\partial_t u^\varepsilon(t)\|^2_{G_2^\varepsilon} + \|\partial_t v^\varepsilon(t)\|^2_{G_2^\varepsilon}$$
$$+ \int_0^t \|\nabla\partial_t U^\varepsilon\|^2_{G_1^\varepsilon} + \int_0^t \|\varepsilon\nabla\partial_t u^\varepsilon\|^2_{G_2^\varepsilon} + \int_0^t \|\varepsilon\nabla\partial_t v^\varepsilon\|^2_{G_2^\varepsilon} \leq C.$$

Proof. Differentiate the system with respect to time and test with the time derivatives of the unknowns; see also Proposition 5 in Hornung *et al.* (1994). □

Most likely, we can perform the homogenization step without that much regularity. It is probably sufficient to ensure that $\partial_t U^\varepsilon \in L^2(S \times G_1^\varepsilon)$, which gives compactness for U^ε. Note, however, that we cannot get compactness for u^ε and v^ε due to the ε^2-scaling in the elliptic terms.

Theorem 5.1. *There exists a unique weak solution to* (P_k^ε).

Proof. The validity of this solvability statement is guaranteed by using standard techniques, for example by cutting off the nonlinearities and using Schauder's fixed-point theorem. We refer the reader to Theorem 1 in Hornung *et al.* (1994) for a precise proof for the case of a quite similar system of partial differential equations. □

It is worth noting that strong solutions to (P_k^ε) (or even classical solutions to the same problem) can in general not be expected. This is due to the presence of mixed boundary conditions for U^ε. Also, the domains are not smooth (rectangles). However, these regularity restrictions will be removed naturally in the limit system.

Proposition 5.4. *There exist extensions* U^ε, u^ε, *and* v^ε *defined on all of* G *such that, at least for a subsequence, the following convergences hold:*

$$U^\varepsilon \rightharpoonup U \qquad \text{weakly in } L^2(S; H^1(G)),$$

$$U^\varepsilon \to U \qquad \text{strongly in } L^2(S \times G),$$

$$U^\varepsilon \overset{2}{\rightharpoonup} U \qquad \text{two-scale},$$

$$\nabla U^\varepsilon \overset{2}{\rightharpoonup} \nabla_x U(x) + \nabla_y U_1(x,y) \qquad \text{two-scale},$$

$$u^\varepsilon \rightharpoonup u \qquad \text{weakly in } L^2(S; H^1(G)),$$

$$u^\varepsilon \overset{2}{\rightharpoonup} u \qquad \text{two-scale},$$

$$\varepsilon \nabla u^\varepsilon \overset{2}{\rightharpoonup} \nabla_y u(x,y) \qquad \text{two-scale},$$

$$v^\varepsilon \rightharpoonup v \qquad \text{weakly in } L^2(S; H^1(G)),$$

$$v^\varepsilon \overset{2}{\rightharpoonup} v \qquad \text{two-scale},$$

$$\varepsilon \nabla v^\varepsilon \overset{2}{\rightharpoonup} \nabla_y v(x,y) \qquad \text{two-scale}.$$

Proof. These convergence results are basically standard; for more details, see Propositions 8 and 9 in Hornung *et al.* (1994). To be able to prove them, first an extension theorem needs to be applied in the same way as reported by Hornung *et al.* (1994) or Hornung and Jäger (1991). Then, as the next and final step, the two-scale convergence results stated above follow from well-known convergence results; see, e.g. Proposition 1.14 given by Allaire (1992). □

5.3 Passage to the Homogenization Limit

Let $\theta := (1-R)/2$. Recall that $\Omega = (0, L)$ and $Y = (0, R)$. We denote by Problem (P_k) the following two-scale system in one space dimension:

Find (U, u, v) satisfying

$$\theta\partial_t U - \theta D\partial_{xx} U = -b(U - u|_{y=R}) \quad x \in \Omega,\ t \in S, \tag{5.19}$$

$$\partial_t u - d_1\partial_{yy} u = -kuv \quad x \in \Omega,\ y \in Y,\ t \in S, \tag{5.20}$$

$$\partial_t v - d_2\partial_{yy} v = -\alpha kuv \quad x \in \Omega,\ y \in Y,\ t \in S, \tag{5.21}$$

with boundary conditions on the microscale

$$d_1\partial_y u(t, x, R) = -b(u(t, x, R) - U(t, x)), \tag{5.22}$$

$$\partial_y u(t, x, 0) = \partial_y v(t, x, 0) = \partial_y v(t, x, R) = 0 \quad x \in \Omega,\ t \in S, \tag{5.23}$$

exterior boundary conditions

$$U(t, 0) = U^{ext}(t), \tag{5.24}$$

$$D\partial_x U(t, R) = 0 \quad t \in S, \tag{5.25}$$

and initial conditions

$$U(0, x) = U_0(x), \tag{5.26}$$

$$u(0, x, y) = u_0(x, y), \tag{5.27}$$

$$v(0, x, y) = v_0(x, y), \quad x \in \Omega,\ y \in Y. \tag{5.28}$$

We formulate now (P_k) in a weak setting as follows: Let $H^1_L(\Omega)$ denote the functions in $H^1(\Omega)$ vanishing at $x = 0$. Define

$$\begin{aligned}
V_1 &:= H^1_L(\Omega),\\
V_2 &:= L^2(\Omega; H^1(Y)),\\
V &:= V_1 \times [V_2]^2,\\
H &:= L^2(\Omega) \times [L^2(\Omega \times Y)]^2,\\
\mathcal{V}_i &:= L^2(S; V_i), \quad i \in \{1, 2\}, \quad \mathcal{V} := L^2(S; V).
\end{aligned}$$

We derive the weak formulation of Problem (P_k) by multiplying equations (5.19)–(5.21) with the test functions $(\phi, \Phi, \psi) \in V$, respectively, and integrating the outcome corresponding to (5.19) over Ω and the outcomes corresponding to (5.20) and to (5.21) over $\Omega \times Y$.

The result is as follows: Find essentially bounded functions $U \in U^{ext} + \mathcal{V}_1$ and $u, v \in \mathcal{V}_2$ that satisfy (5.26)–(5.28) and

$$\begin{aligned}
&\frac{\mathrm{d}}{\mathrm{d}t}\int_\Omega \theta U\phi + \int_\Omega \theta D\partial_x U \partial_x \phi + \int_\Omega b(U - u|_{y=R})(\phi - \Phi|_{y=R}) + \frac{\mathrm{d}}{\mathrm{d}t}\int_\Omega\int_Y u\Phi\\
&\quad + \int_\Omega\int_Y d_1 \partial_y u \partial_y \Phi + \int_\Omega\int_Y kuv\Phi = 0 \quad \text{for all } (\phi, \Phi) \in V_1 \times V_2,
\end{aligned} \tag{5.29}$$

$$\frac{\mathrm{d}}{\mathrm{d}t}\int_\Omega\int_Y v\Psi + \int_\Omega\int_Y d_2 \partial_y v \partial_y \Psi + \int_\Omega\int_Y \alpha kuv\Psi = 0 \quad \text{for all } \Psi \in V_2. \tag{5.30}$$

It is worth noting that, in this formulation, the essential boundedness of the involved functions is only needed in order to give meaning to the nonlinear reaction terms.

Theorem 5.2. *The limit functions U, u, and v satisfy the two-scale problem (P_k) in the weak sense.*

Proof.
Step 1: The limit system in space dimension 2. Relying on the *a priori* estimates stated in Propositions 5.1–5.3, we can apply the same monotonicity trick as done by Hornung *et al.* (1994) (Proposition 12) to pass to the limit $\varepsilon \to 0$ via two-scale convergence. In the strong formulation, the limit

equations are as follows:

$$|Z_1|\partial_t U - \nabla\cdot(\mathbb{D}\nabla U) = -\int_\Gamma b(U-u) \quad x\in G,$$
$$\partial_t u - d_1\Delta u = -kuv \quad x\in G,\ y\in Z_2,$$
$$\partial_t v - d_2\Delta v = -\alpha kuv \quad x\in G,\ y\in Z_2.$$

The effective diffusion tensor is given by

$$\mathbb{D} := D\int_{Z_1}\begin{pmatrix} 1+\partial_{y_1}\zeta_1 & \partial_{y_1}\zeta_2 \\ \partial_{y_2}\zeta_1 & 1+\partial_{y_2}\zeta_2 \end{pmatrix} dy,$$

where ζ_1, ζ_2 are Z-periodic solutions of

$$\Delta_y\zeta_j = 0 \quad y\in Z_1 \tag{5.31}$$
$$-\nabla_y\zeta\cdot\nu = e_j\cdot\nu \quad y\in\Gamma. \tag{5.32}$$

Moreover, the limit functions are subject to the same L^∞-bounds as that from Proposition 5.1.

Step 2: Dimension reduction in the microscopic problem. Due to the symmetry of the geometry and data, it follows immediately that u and v are independent of y_1, i.e. $u = u(t,x,y_2)$ and $v = v(t,x,y_2)$. Moreover, it is sufficient to consider only $y_2 \in (0,R) = Y$. From that, we obtain (5.20) and (5.21). The macroscopic problem is still two-dimensional, but it simplifies to

$$(1-R)\partial_t U - \nabla\cdot(\mathbb{D}\nabla U) = -2b(U-u|_{y=R}), \qquad x\in G.$$

Step 3: Dimension reduction in the macroscopic problem. The solutions of (5.31)–(5.32) are given by

$$\zeta_2(y) = -y_2 + c_1 \quad \text{and} \quad \zeta_1(y) = c_1.$$

This gives

$$\partial_{y_1}\zeta_2 = 0 = \partial_{y_1}\zeta_1 = \partial_{y_2}\zeta_1 \quad \text{and} \quad \partial_{y_2}\zeta_2 = -1.$$

Hence, we get

$$\mathbb{D} = D\int_{Z_1}\begin{pmatrix} 1 & 0 \\ 0 & 0 \end{pmatrix} dy = \begin{pmatrix} D(1-R) & 0 \\ 0 & 0 \end{pmatrix}.$$

The explicit structure of $\mathbb{D}$ points out that diffusion is restricted to the x_1-direction. It follows by symmetry that $u = u(t, x_1)$ is actually independent of x_2 and satisfies (5.20). □

Remark 5.1. The positivity and L^∞-bounds of the weak solutions are preserved during the homogenization process. On the other hand, they can also be proved directly for (P_k).

Lemma 5.1. *It holds that*

$$\|kuv\|_{L^1(S\times\Omega\times Y)} \leq \frac{1}{\alpha}\|v_0\|_{L^1(\Omega\times Y)}. \tag{5.33}$$

Proof. Integrating (5.21) along $S \times \Omega \times Y$, we directly obtain

$$\begin{aligned}
\alpha \int_0^T \int_\Omega \int_Y kuv &= -\int_0^T \int_\Omega \int_Y \partial_t v - d_2 \partial_{yy} v \\
&= -\underbrace{\int_\Omega \int_Y v(T,x,y)}_{\geq 0} + \int_\Omega \int_Y v_0(x,y) + \underbrace{\int_0^t \int_\Omega \int_{\partial Y} d_2 \partial_y v}_{=0} \\
&\leq \int_\Omega \int_Y v_0(x,y).
\end{aligned}$$

□

Lemma 5.2 (Energy estimates independent of k). *There exists a constant $c > 0$, which is independent of k, such that the following estimates hold:*

$$\max_{t\in S} \int_{\Omega\times Y} v^2 + \int_0^T \int_\Omega \int_Y |\partial_y v|^2 \leq c \tag{5.34}$$

$$\max_{t\in S} \left(\int_\Omega U^2 + \int_{\Omega\times Y} u^2 \right) + \int_0^T \int_\Omega |\partial_x U|^2 + \int_0^T \int_\Omega \int_Y |\partial_y u|^2 \leq c.$$

Proof. Testing (5.21) by v, we obtain

$$\int_0^T \frac{1}{2}\frac{d}{dt} \int_\Omega \int_Y v^2 + d_2 \int_0^T \int_\Omega \int_Y |\partial_y v|^2 + \alpha k \underbrace{\int_0^t \int_\Omega \int_Y uv^2}_{\geq 0} = 0. \tag{5.35}$$

This leads to the inequality

$$\frac{1}{2}\int_\Omega\int_Y v(t)^2 + d_2\int_0^t\int_\Omega\int_Y |\partial_y v|^2 \le \frac{1}{2}\int_\Omega\int_Y v_0^2,$$

which finally guarantees (5.34). Furthermore, testing (5.19) by $U - U^{ext}$ gives

$$\frac{\theta}{2}\frac{d}{dt}\int_\Omega U^2 - \theta\int_\Omega U\partial_t U^{ext} + \theta D\int_\Omega(\partial_x U)^2 = -b\int_\Omega(U - u|_{y=R})(U - U^{ext}), \tag{5.36}$$

and testing (5.20) by $u - U^{ext}$ gives

$$\begin{aligned}\frac{1}{2}\frac{d}{dt}\int_\Omega\int_Y u^2 - \int_\Omega\int_Y u\partial_t U^{ext} + d_1\int_\Omega\int_Y(\partial_y u)^2 \\ - b\int_\Omega(U - u|_{y=R})(u - U^{ext}) = -\int_\Omega\int_Y kuv(u - U^{ext}).\end{aligned} \tag{5.37}$$

Adding together (5.36) and (5.37) and integrating the result from 0 to t yields with standard estimates

$$\begin{aligned}\theta\int_\Omega U(t)^2 + 2\theta D\int_0^t\int_\Omega(\partial_x U)^2 + \int_\Omega\int_Y u(t)^2 + 2d_1\int_0^t\int_\Omega\int_Y(\partial_y u)^2 \\ \le \theta\int_\Omega U_0^2 + \int_\Omega\int_Y u_0^2 + \|\partial_t U^{ext}\|_S^2 + \theta^2\int_0^t \|U\|_\Omega^2 + \int_0^t \|u\|_{\Omega\times Y}^2 \\ + \|U^{ext}\|_\infty\int_0^t\int_\Omega\int_Y kuv.\end{aligned}$$

Grönwall's inequality and Lemma 5.1 lead to (5.34). □

Denoting by w the expression $u - \frac{1}{\alpha}v$, we see that

$$\partial_t w = d_1\Delta_y u - \frac{d_2}{\alpha}\Delta_y v. \tag{5.38}$$

Proposition 5.5. *There exists a constant c_1, which is independent of k, such that*

$$\|\partial_t U\|_{\mathcal{V}_1'} + \|\partial_t w\|_{\mathcal{V}_2'} \le c_1. \tag{5.39}$$

Proof. Take $(\phi, \xi) \in \mathcal{V}_1 \times \mathcal{V}_2$ arbitrarily. Testing (5.38) with ξ gives

$$\int_0^T \langle \partial_t w, \xi \rangle = -d_1 \int_0^T \int_\Omega \int_Y \partial_y u \partial_y \xi + \frac{d_2}{\alpha} \int_0^T \int_\Omega \int_Y \partial_y v \partial_y \xi - b \int_0^T \int_\Omega (u(t,x,R) - U(t,x)\xi(t,x,R), \tag{5.40}$$

and testing (5.19) with ϕ yields

$$\theta \int_0^T \langle \partial_t U, \phi \rangle = -\theta \int_0^T \int_\Omega \partial_x U \partial_x \phi - b \int_0^T \int_\Omega (U(x,t) - u(t,x,R))\phi(t,x). \tag{5.41}$$

Using the Cauchy–Schwarz and Hölder inequalities, as well as the k-independent energy estimates stated in Lemma 5.2, we get the estimate

$$\begin{aligned} \left| \int_0^T \langle \partial_t w, \xi \rangle \right| &\leq c_0 \left(||\partial_y u||_{L^2(S\times\Omega\times Y)} + ||\partial_y v||_{L^2(S\times\Omega\times Y)} \right) ||\partial_y \xi||_{L^2(S\times\Omega\times Y)} \\ &\quad + bC_M \int_0^T \int_\Omega ||\xi||_{H^1(Y)} \\ &\leq c||\xi||_{L^2(S\times\Omega\times H^1(Y))}, \end{aligned} \tag{5.42}$$

where the constant c is independent of k. Therefore, it holds that

$$||\partial_t w||_{L^2(S\times\Omega;(H^1(Y))^\star)} \leq c < \infty.$$

Proceeding in a similar way, we also obtain that

$$||\partial_t U||_{L^2(S;H^1_L(\Omega)^\star)} \leq c < \infty.$$

Finally, adding the latter two estimates, we obtain (5.39). □

The estimate (5.39) encourages us to use the compactness criterion provided via the following Lions–Aubin lemma (as stated in a slightly different form in Lemma 7.7, the original reference work being that by Aubin (1963)).

Lemma 5.3 (Lions–Aubin). *Let $B_0 \hookrightarrow B \hookrightarrow B_1$ be Banach spaces such that B_0 and B_1 are reflexive and the embedding $B_0 \hookrightarrow B$ is compact. Fix $p, q > 0$, and let*

$$W = \left\{ z \in L^p(S; B_0) : \frac{dz}{dt} \in L^q(S; B_1) \right\},$$

with

$$||z||_W := ||z||_{L^p(S;B_0)} + ||\partial_t z||_{L^q(S;B_1)}.$$

Then, $W \hookrightarrow\hookrightarrow L^p(S; B)$.

Since the embedding $V_2 \hookrightarrow L^2(\Omega \times Y)$ is not compact, Lemma 5.3 cannot be applied for the whole two-scale system. However, it is applicable for the choice $(B_0, B, B_1) := (H^1(Y), L^2(Y), H^1(Y)^\star)$. This implies that

$$W := \{w \in L^2(S, H^1(Y)) : \ \partial_t w \in L^2(S; H^1(Y)^\star)\} \hookrightarrow\hookrightarrow L^2(S \times Y).$$

Depending on the choice of the micro–macro coupling in the two-scale problem, additional regularity of the microscopic unknowns with respect to the macroscopic variable x is sometimes needed to get the desired compactness for the problem stated at the microscopic level. This can be done by differentiating the microscopic system with respect to $x \in \Omega$ and then employing a characterization of the spaces $H^1(\Omega, B)$ in terms of translation estimates (cf. Lemmas 7.23 and 7.24 in Gilbarg and Trudinger (1998) for the case of $B = \mathbb{R}$).

5.4 Passage to the Fast-Reaction Limit

We deduce from Lemma 5.2 that the sequence (U_k) is bounded in $L^2(S; H^1(\Omega))$ and in $L^\infty(S; L^2(\Omega))$ and that the sequences (u_k) and (v_k) are bounded in $\mathcal{V}_2$ and in $L^\infty(S; L^2(\Omega \times Y))$. Thus, there exist subsequences of (U_k), (u_k), and (v_k), which we denote in the same way as the full respective sequence, and limit functions $(\overline{U}, \overline{u}) \in \mathcal{V}_1$ and $\overline{v} \in \mathcal{V}_2$ such that

$$0 \le \overline{U}, \overline{u}, \overline{v} \le C_M \quad \text{a.e.}$$

and

$$U_k \rightharpoonup \overline{U} \qquad \text{weakly in } L^2(S; H^1(\Omega)), \tag{5.43}$$

$$U_k(T) \rightharpoonup \overline{\kappa}_1 \qquad \text{weakly in } L^2(\Omega), \tag{5.44}$$

$$u_k \rightharpoonup \overline{u}, \ v_k \overset{\text{w}}{\rightharpoonup} \overline{v} \qquad \text{weakly in } \mathcal{V}_2, \tag{5.45}$$

$$u_k(T) \rightharpoonup \overline{\kappa}_2, \ v_k(T) \rightharpoonup \overline{\kappa}_3 \qquad \text{weakly in } L^2(\Omega \times Y). \tag{5.46}$$

Results from the previous section provide, for a.e. $x \in \Omega$, the needed relative compactness of $w_k(\cdot, x, \cdot)$ in $L^2(S \times Y)$, which ensures that

$$w_k = u_k - \frac{v_k}{\alpha} \to w \quad \text{in } L^2(S \times Y) \quad \text{as } k \to \infty. \tag{5.47}$$

It follows for a subsequence that $w_k \to w$ for almost every $(t, x, y) \in S \times \Omega \times Y$ and, therefore, by the uniform boundedness of u_k and v_k,

$$w_k \to w \quad \text{in } L^p(S \times \Omega \times Y) \text{ as } k \to \infty \tag{5.48}$$

for any $p \geq 1$.

Moreover, from the estimate in Lemma 5.1, we deduce that

$$u_k v_k \to 0 \quad \text{in } L^1(S \times \Omega \times Y) \text{ as } k \to \infty.$$

Lemma 5.4. *The following statements hold:*

(i) *The subsequences u_k and v_k are such that*

$$u_k \to w^+ \quad \textit{and} \quad v_k \to \alpha w^- \quad \textit{as } k \to \infty$$

in $L^1(S \times \Omega \times Y)$ and a.e. in $S \times \Omega \times Y$;

(ii) $\bar{u} = w^+$ *and* $\bar{v} = \alpha w^-$.

Proof. Since (ii) is a straightforward consequence of (i), we only show a proof for (i). To this end, we follow the lines of the proof of Lemma 3.1 from Dancer *et al.* (1999). Recall that for a.e. $(t, x, y) \in S \times \Omega \times Y$, we have $w_k = \left(u_k - \frac{v_k}{\alpha}\right)(t, x, y) \to w(t, x, y)$ and $(u_k v_k)(t, x, y) \to 0$ as $k \to \infty$.

We distinguish between three distinct cases:

(1) $w(t, x, y) > 0$: In this case, for all $\eta > 0$, there exists a rank $k_\eta \in \mathbb{N}$ such that for all $k \geq k_\eta$, we have $|u_k - \frac{v_k}{\alpha} - w| \leq \eta$. In particular, we obtain $u_k - \frac{v_k}{\alpha} \geq w - \eta$. Choosing now $\eta := \frac{w}{2}$, we use the positivity of v_k to get $u_k \geq \frac{w}{2}$.
(2) $w(t, x, y) < 0$: We proceed analogously as in case (1).
(3) $w(t, x, y) = 0$: We argue now via *reductio ad absurdum*. Assume that there exists a subsequence $u_k \to \rho > 0$. Then, the separation in space of the reactants forces $v_k \to 0$ as $k \to \infty$, and hence, $u_k - \frac{v_k}{\alpha} \to \rho$. On the other hand, we recall that $u_k - \frac{v_k}{\alpha} \to 0$ as $k \to \infty$, which contradicts our sign assumption on ρ.

Since all mentioned subsequences are essentially bounded, we obtain by Lebesgue's theorem of dominated convergence that these subsequences are also L^p convergent for all $p \geq 1$. $\square$

In what follows, we want to derive both weak and strong formulations for the limit problem satisfied by w and $\overline{U}$.

Proposition 5.6. *The pair of functions* $(\overline{U}, w)$ *defined via* (5.43) *and* (5.47) *satisfies problem* (P), *which is*

$$-\int_0^T\int_\Omega \theta\overline{U}\partial_t\phi - \int_\Omega \theta U_0\phi(0) + \int_0^T\int_\Omega \theta D\partial_x\overline{U}\partial_x\phi - \int_0^T\int_\Omega\int_Y w\partial_t\Phi$$
$$- \int_\Omega\int_Y \left(u_0 - \frac{v_0}{\alpha}\right)\Phi(0) + \int_0^T\int_\Omega\int_Y\left(d_1\partial_y\overline{u} - \frac{d_2}{\alpha}\partial_y\overline{v}\right)\partial_y\Phi$$
$$= -b\int_0^T (\bar{U} - \bar{u}|_{y=R})(\phi - \Phi|_{y=R}) \tag{5.49}$$

for all test functions $\phi \in C^1(S; H^1_L(\Omega))$ *and* $\Phi \in C^1(S; L^2(\Omega; H^1(Y)))$ *such that* $\phi(T) = \Phi(T) = 0$.

Proof. We multiply the strong formulation of the w_k-problem by the test functions ϕ and Φ as stated above. Integration by parts gives

$$-\int_0^T\int_\Omega \theta U_k\partial_t\phi - \int_\Omega \theta U_0\phi(0) + \int_0^T\int_\Omega \theta D\partial_x U_k\partial_x\phi - \int_0^T\int_\Omega\int_Y w_k\partial_t\Phi$$
$$- \int_\Omega\int_Y \left(u_0 - \frac{v_0}{\alpha}\right)\Phi(0) + \int_0^T\int_\Omega\int_Y\left(d_1\partial_y u_k - \frac{d_2}{\alpha}\partial_y v_k\right)\partial_y\Phi$$
$$= -b\int_0^T\int_\Omega (U_k - u_k|_{y=R}(\phi - \Phi|_{y=R}).$$

Relying on the weak convergences (5.43)–(5.46), we pass in each term on the left-hand side to its corresponding limit as $k \to \infty$. Thus, we obtain in the limit (5.49). □

Theorem 5.3. *Assume* $d_1 = d_2/\alpha$. *Then, there exists at most a couple* $(\bar{U}, w)$ *satisfying the limit problem* (P) *in a weak sense.*

Proof. Since problem (P) is linear, we choose $w_0 = 0, U^{ext} = 0$, and $\bar{U}_0 = 0$, and we discuss the uniqueness problem for the case of $b \in (0, 1]$.

Testing in the weak formulation of problem (P) by $\bar{U}$ and w, we obtain

$$\frac{\theta}{2}\frac{d}{dt}\int_\Omega |\bar{U}|^2 + \theta D \int_\Omega |\partial_x \bar{U}|^2 = -b\int_\Omega (\bar{U} - w^+|_{y=R})\bar{U}, \tag{5.50}$$

$$\frac{1}{2}\frac{d}{dt}\int_\Omega\int_Y |w|^2 + d_1 \int_\Omega\int_Y |\partial_y w|^2 = b\int_\Omega (\bar{U} - w^+|_{y=R})w. \tag{5.51}$$

Summing up (5.50) and (5.51), we obtain

$$\begin{aligned}\frac{\theta}{2}\frac{d}{dt}\int_\Omega |\bar{U}|^2 + \frac{1}{2}\frac{d}{dt}\int_\Omega\int_Y |w|^2 + \theta D\int_\Omega |\partial_x\bar{U}|^2 + d_1\int_\Omega\int_Y |\partial_y w|^2 \\ = -b\int_\Omega (\bar{U} - w^+)(\bar{U} - w) \\ = -b\int_\Omega (\bar{U} - w^+)^2 \underbrace{-b\int_\Omega (\bar{U} - w^+)w^-}_{\leq 0} \\ \leq 0 \end{aligned} \tag{5.52}$$

Integrating now (5.52) along $(0,t)$, we have

$$\frac{1}{2}\int_\Omega \theta\bar{U}(t)^2 + \frac{1}{2}\int_\Omega\int_Y w(t)^2 \leq 0,$$

which concludes the proof of the theorem. □

As the next step, we wish to recover the strong formulation of problem (P). Note, however, that in general the second derivative of w may not exist. Choosing $\phi \equiv 0$ and $\Phi \in C_0^\infty(S\times\Omega\times Y)$ and partial integration back yields

$$\partial_t w - d_1\partial_{yy}\overline{u} - \frac{d_2}{\alpha}\partial_{yy}\overline{v} = 0 \quad \text{in } Y.$$

Testing the result with some Φ being nonzero at $t = 0$ and $y = 0$, respectively, yields the initial condition $w(0) = u_0 - \frac{1}{\alpha}v_0$ and the boundary condition

$$d_1\partial_y\overline{u}|_{y=0} - \frac{d_2}{\alpha}\partial_y\overline{v}|_{y=0} = 0.$$

Furthermore, testing with a function which is not vanishing at $y = R$, we obtain

$$-d_1\partial_y\bar{u} + \frac{d_2}{\alpha}\partial_y\bar{v} = -b(\bar{u}(t,x,R) - \bar{U}(t,x)).$$

Assume for a moment that we have

$$w(t, x, R) > 0 \quad \text{for all } t \text{ and } x. \tag{5.53}$$

Then, by the continuity of w in $y = R$, it follows that

$$\partial_y \bar{v}|_{y=R} = 0. \tag{5.54}$$

Note that (5.53) can be removed if instead greater regularity of the initial data u_0, v_0, and U_0 is provided. For related remarks, see Remark 1.1. and Theorem 1.3 in DiBenedetto (1993), p. 43, Lemma 3.5 in Dancer *et al.* (1999), or Theorem 1.1 in Friedman and Tzavaras (1987).

Letting now $\phi \in C_0^\infty(S \times \Omega)$ and $\Phi \equiv 0$, the integration by parts yields

$$\theta \partial_t \overline{U} - \theta D \partial_{xx} \overline{U} = -b(\bar{U} - \bar{u}|_{y=R}) \quad \text{in } \Omega.$$

Testing afterward with a ϕ not vanishing at $x = L$, we recover the boundary condition $\partial_y \overline{U}|_{x=L} = 0$. Additionally, we also recover the initial condition $\overline{U}(0) = \overline{U}_0$ by the standard technique (see, e.g. Evans (2015) for an example for the derivation of $\overline{U}(0) = \overline{U}_0$).

Finally, we arrive at the following strong formulation for (P). Define $d(w) := d_1$ if $w \geq 0$ and $d(w) := \frac{d_2}{\alpha}$ otherwise. Note that $d(w)\partial_{yy} w = \partial_y(d(w)\partial_y w)$.

$$\begin{aligned}
\theta \partial_t \overline{U} - \theta D \partial_{xx} \overline{U} &= -b(\bar{U} - w^+|_{y=R}) \quad x \in \Omega,\ t \in S, \\
\partial_t w - \partial_y(d(w)\partial_y w) &= 0 \quad x \in \Omega,\ y \in Y,\ t \in S, \\
-d(w)\partial_y w(t, x, R) &= b(w^+(t, x, R) - \overline{U}(t, x)) \quad x \in \Omega,\ t \in S, \\
\partial_y w^-(t, x, R) &= 0 \quad x \in \Omega,\ t \in S, \\
\partial_y w(t, x, 0) &= 0 \quad x \in \Omega,\ t \in S, \\
\overline{U}(t, 0) &= U^{ext}(t) \quad t \in S, \\
\partial_x \overline{U}(t, L) &= 0 \quad t \in S,
\end{aligned}$$

with the initial conditions $\overline{U}(0) = U_0$ and $w(0) = u_0 - \frac{1}{\alpha} v_0$.

5.5 Discovering Microscopic Free Boundaries

Under sufficient regularity assumptions on the involved functions and geometry, the limit problem (P) can be written as a two-scale free-boundary problem with the free boundary concentrated in the microscale, as we conjectured for the case of a related problem with matched micro–macro boundary conditions; see Meier and Muntean (2008) (Section 3.2).

We denote by Σ the non-empty support of $\overline{v}$, i.e. we take

$$\Sigma = \{\{t\} \times \{x\} \times (0, s(t,x)) : \ t \in S, \ x \in \Omega\}, \tag{\textbf{G}}$$

where $s(t,x) \in (0,R)$ denotes the point separating the support of $\bar{u}(\cdot,x,\cdot)$ from that of $\bar{v}(\cdot,x,\cdot)$ for a.e. $x \in \Omega$. We refer to $s(t,x)$ as the position of the *a priori* unknown free boundary at time $t \in S$, corresponding to the point $x \in \Omega$. Moreover, we assume that $\overline{u} > 0$ outside $\overline{\Sigma}$, i.e. the interface is in fact sharp and the reaction in each local cell concentrates on the point $y = s(t,x)$.

The initial position s_0 of the free boundary is immediately recovered since the initial values of the microscopic concentrations – the solutions of the (P_k) problem – have initially disjoint supports, i.e. there is a function $s_0 : \Omega \to (0,R)$ such that $w < 0$ in $(0, s_0(x))$ and $w > 0$ in $(s_0(x), R)$. It is worth mentioning at this point that in Remark 1 from Seidman (2009), the choice of s_0 is related to an upper estimation of the size of the initial transient layer for a reaction-diffusion system with fast reaction.

To simplify matters, we assume that

$$\bar{U}, \bar{u}, \bar{v} \text{ and } s \text{ are sufficiently smooth.} \tag{\textbf{R}}$$

If assumption (**R**) is satisfied, then problem (P) can be reformulated as follows:

Find $(\bar{U}, \bar{u}, \bar{v}, s)$ satisfying the model equations

$$\theta \partial_t \bar{U} - \theta D \partial_{xx} \bar{U} = -d_1 \partial_y \bar{u}(t,x,R) \quad \text{in } \Omega, \tag{5.55}$$

$$\partial_t \bar{u} - d_1 \partial_{yy} \bar{u} = 0 \quad \text{in } (s(t,x), R), \tag{5.56}$$

$$\partial_t \bar{v} - d_2 \partial_{yy} \bar{v} = 0 \quad \text{in } (0, s(t,x)), \tag{5.57}$$

the boundary conditions

$$d_1 \partial_y \bar{u}(t,x,R) = -b(\bar{u}(t,x,R) - \bar{U}(t,x)), \tag{5.58}$$

$$-d_2 \partial_y \bar{v}(t,x,0) = 0, \tag{5.59}$$

$$-d_1\partial_y\bar{u}(t,x,s(t,x)) = \frac{d_2}{\alpha}\partial_y\bar{v}(t,x,s(t,x)), \tag{5.60}$$

$$\bar{u}(t,x,s(t,x)) = 0 = \bar{v}(t,x,s(t,x)), \tag{5.61}$$

and the initial conditions

$$\bar{U}(0,x) = U_0(x), \bar{u}(0,x,y) = u_0(x,y), \tag{5.62}$$

$$\bar{v}(0,x,y) = v_0(x,y), s(0,x) = s_0(x) \tag{5.63}$$

for all $(x,y) \in \Omega \times Y$ and $t \in S$.

We refer to (5.55)–(5.63) as (FBP).

The presence of two spatial scales leads to a non-standard free-boundary problem (FBP). If we would ignore for a moment the macroscopic equation (5.55), then the remaining equations would constitute the classical two-phase Stefan problem with zero latent heat and instantaneous reaction. We refer the reader to Evans (1982) for the existence, uniqueness, and regularity study of such a problem posed in one-space dimension and to Cannon and Hill (1970) for the analysis of its multi-dimensional counterpart. By assuming that $s(t,x) > 0$, we exclude the case where the moving boundary reaches the left end of the interval. Note that in this case, the partial differential equation (5.57) and the boundary conditions (5.59) and (5.60) lose their meaning.

Now, we show that the production term by reaction tends to concentrate on a measure localized within the microstructure. To do so, we adapt some ideas from Hilhorst *et al.* (2012). Let μ_k denote the expression $-ku_kv_k$.

Theorem 5.4 (Convergence of the reaction term to a measure). *Assume that (R) and (G) hold. Then, μ_k converges to the Radon measure μ, given by*

$$\mu(t,x,y) = \frac{1}{2}\left(-d_1\partial_y u(t,x,y) + d_2\partial_y v(t,x,y)\right)\delta(y - s(t,x)), \tag{5.64}$$

where $(t,x,y) \in S \times \Omega \times Y$.

Proof. Let us take $\psi \in C_0^\infty(S \times Y)$. Integrating by parts, we get

$$\begin{aligned}\int_0^T \int_Y \mu_k \psi dydt &= \frac{1}{\alpha} \int_0^T \int_Y v_{kt}\psi - d_2 \partial_{yy}^2 v_k \psi \\ &= -\frac{1}{\alpha} \int_0^T \int_Y v_k \partial_t \psi + \frac{d_2}{\alpha} \int_0^T \int_Y \partial_y v_k \partial_y \psi \\ &= \int_0^T \int_Y u_{kt}\psi - d_1 \partial_{yy} u_k \psi \\ &= \int_0^T \int_Y -u_k \partial_t \psi + d_1 \int_0^T \int_Y \partial_y u_k \partial_y \psi.\end{aligned}$$

Passing to the limit $k \to \infty$ in the last expression yields

$$\int_0^T \int_Y \mu\psi dydt = -\frac{1}{\alpha} \int_0^T \int_0^{s(t)} \bar{v}\partial_t \psi + \frac{d_2}{\alpha} \int_0^T \int_0^{s(t)} \partial_y \bar{v} \partial_y \psi \quad (5.65)$$

$$= \int_0^T \int_{s(t)}^R -\bar{u}\partial_t \psi + d_1 \int_0^T \int_{s(t)}^R \partial_y \bar{u} \partial_y \psi. \quad (5.66)$$

Putting now in (5.65) and (5.66) the derivatives on $\bar{u}$ and $\bar{v}$ and using the limit equations (5.56) and (5.57) for $\bar{u}$ and $\bar{v}$, we obtain

$$\begin{aligned}\int_0^T \int_Y \mu\psi &= \int_0^T -d_1 \partial_y \bar{u}(t, x, s(t, x))\psi(t, s(t, x)) \\ &= \int_0^T \frac{d_2}{\alpha} \partial_y \bar{v}(t, x, s(t, x))\psi(t, s(t, x)).\end{aligned}$$

This argument proves (5.64). □

5.6 Discussion

We refer the reader to two other interesting mathematical settings, where physical scenarios, namely, linear elasticity and diffusion, are described by evolution equations within a macroscopic domain. These are coupled with evolution equations governing freely moving interfaces, which are posed at a microscopic spatial scale, as discussed by Aiki *et al.* (2021) and Kumazaki and Muntean (2025).

Such models offer a very rich insight from a multiscale modeling point of view, but they are challenging to handle numerically, especially in higher spatial dimensions. A simple, yet computationally efficient case is discussed by Lakkis *et al.* (2024).

5.7 Exercises

Exercise 5.7.1. Prove Proposition 5.1.

Exercise 5.7.2. Prove Proposition 5.2.

Exercise 5.7.3. Prove Proposition 5.3.

Exercise 5.7.4. Prove the existence of at least a weak solution to (FBP), that is, for the system (5.55)–(5.63).

Exercise 5.7.5. Solve numerically the problem (FBP) for a selection of model parameters of your choice.

Exercise 5.7.6. Formulate a reaction-diffusion system involving three chemical species which leads (formally) in the fast-reaction limit to a boundary-value problem involving two distinct freely moving boundaries.

5.8 Solutions

Solution 5.7.1. Here, one needs to prove a weak maximum principle for a linear parabolic problem. Among other possible solution strategies, both Stampacchia's technique and the method of lower and upper solutions are applicable.

Solution 5.7.2. This exercise is about proving a standard energy estimate for a linear parabolic problem. The reader can find a couple of solved exercises of this type in Evans (2015).

Solution 5.7.3. Same hint as for the the previous exercise.

Solution 5.7.4. It is worth noting that this micro–macro boundary-value problem has a freely moving unknown interface present in each of the microscopic domains. This free interface is driven by a two-phase Stefan problem with zero latent heat, and precisely this aspect makes the question challenging. In this class of problems, proving that the speed of the interface is finite is not easy.

Solution 5.7.6. The easiest way is to cook up a reaction-diffusion system with two exploding parameters multiplying conveniently written reaction rates. Alternatively, one may think of a phase-field system with two independent phase-field indicators so that the result of the sharp-interface asymptotics in the proposed model coincides with what one would get via fast-reaction asymptotics in a more standard (semi-linear) reaction-diffusion setting. Useful ideas in this context can be found, for instance, in the works of Nishiura (2002) and Seidman (2009) and references cited therein.

Chapter 6

Finite Element Codes in FEniCS for a Linear Elliptic Equation Posed in a Perforated Domain and for Its Homogenized Version

In this chapter, we guide the reader through the numerical approximation of both the weak solution to a microscopic problem and the weak solution to the corresponding macroscopic problem, whose structure is derived using periodic homogenization techniques, as discussed in the previous chapters. To clarify the context, we focus on a linear elliptic equation posed in a perforated domain and its homogenized counterpart. To approximate numerically the wanted weak solutions, we make use of FEniCS; see https://fenicsproject.org. The FEniCS Project is a collection of free and open-source software components aimed at providing programmers with easy access to approximate solutions of differential equations. Such a computational framework is valuable in the context of homogenization from at least two perspectives: First, it helps verify the correctness of the asymptotics; second, it serves as a tool to explore how microscopic effects (such as geometry, parameters, and model components) influence macroscopic observables. This information is essential for collaborating with engineers and scientists interested in the results of the homogenization limiting process.

6.1 Starting Point

We plan to solve numerically a linear stationary reaction-diffusion problem posed in a perforated domain, as well as its macroscopic limit, which is obtained either as the result of two-scale convergence arguments or via two-scale formal asymptotic homogenization. We begin with the presentation of the microscopic model equation. Then, we solve this object numerically for a few different choices of perforated domains obtained for various concrete selections of sufficiently small values of $\varepsilon > 0$. As the next step, we formulate the corresponding macroscopic model equations, and finally, we solve them numerically as well. This last step requires solving the cell problem numerically and computing the values of the effective coefficients.

In this chapter, we rely on FEniCS for performing the computations. The reader should note that there are many other alternatives. The first one to mention would be FreeFEM++ (see https://freefem.org). Other tools could be, for instance, MATLAB (see Abdulle and Nonnenmacher, 2009), `SfePy` (see https://sfepy.org/doc-devel/index.html), `HomPy` (see Ozdilek *et al.*, 2024), and references cited therein, or perhaps using `Julia`. We encourage the reader to choose the computational platform and/or the programming language that best fits their interests and skills.

6.2 Formulation of the Microscopic Problem

We illustrate the main ideas using the following microscopic problem (P_ε):

Find $u_\varepsilon \in H^1(\Omega_\varepsilon; \Gamma_D)$ such that

$$\operatorname{div}(-a_\varepsilon(x)\nabla u_\varepsilon(x)) = -\sqrt{\varepsilon} u_\varepsilon(x) + f_\varepsilon(x) \quad \text{for } x \in \Omega_\varepsilon, \tag{6.1}$$

$$-a_\varepsilon(x)\nabla u_\varepsilon(x) \cdot n_\varepsilon(x) = \varepsilon u_\varepsilon(x) \quad \text{for } x \in \Gamma_\varepsilon^N, \tag{6.2}$$

$$u_\varepsilon(x) = 0 \quad \text{for } x \in \Gamma^D. \tag{6.3}$$

We assume the following restrictions on the data and parameters:

(H1) The matrix A satisfies the coercivity condition: There exists $\alpha, \beta \in (0, \infty)$ independent of ε such that

$$\alpha||\xi||^2_{\mathbb{R}^2} \leq \sum_{i,j=1}^{2} A_{ij}(y)\xi_i\xi_j \leq \beta||\xi||^2_{\mathbb{R}^2}$$

holds for a.e. $y \in Y$ and for all $\xi := (\xi_1, \xi_2) \in \mathbb{R}^2$;

(H2) $A_{ij} \in L^\infty(Y)$ for all $i, j \in \{1, 2\}$, A symmetric, and Y-periodic;
(H3) $f_\varepsilon \in L^2(\Omega_\varepsilon)$ is uniformly bounded, and we assume the structure $f_\varepsilon(x) = f_0(x, \frac{x}{\varepsilon})$ such that $f_0(x, \cdot)$ is continuous in the second variable for any Ω_ε;
(H4) Ω is a rectangle in $\mathbb{R}^2$, while Y is a square with unit area.

As specific choices for the involved parameter functions, we set

$$f_\varepsilon(x) := \sin 2\pi x_1 \sin 2\pi x_2 + \varepsilon \sin \frac{2\pi x_1}{\varepsilon} \cos \frac{2\pi x_2}{\varepsilon}$$

and

$$a_\varepsilon(x) := \begin{pmatrix} 2 + \sin \frac{2\pi x_1}{\varepsilon} \sin \frac{2\pi x_2}{\varepsilon} & 0 \\ 0 & 2 + \sin \frac{2\pi x_1}{\varepsilon} \cos \frac{2\pi x_2}{\varepsilon} \end{pmatrix}$$

for any $x = (x_1, x_2) \in \Omega$. As we wish to employ the finite element method (FEM) to solve numerically this problem and its macroscopic version, a few more specific restrictions need to be added to the list (H1)–(H4). They are mainly concerned with the discretization of Ω_ε and with the choice of basis in the finite-dimensional finite element spaces used in the approximation of u_ε in $H^1(\Omega_\varepsilon; \Gamma_D)$. We omit showing the details here and refer the reader to a standard numerical analysis textbook that addresses this matter; see e.g. Larsson and Thomée (2003).

The microscopic geometry is constructed by covering the target domain Ω with a pavement of scaled squares. Each of the squares has a circular perforation (disk) placed in the center. The resulting set is denoted by Ω_ε; the description of the perforated geometry is borrowed from Section 7.1. We show in Figure 6.1 a patch within the domain Ω_ε. We refer the reader to Section 4.2 for the details of the construction of the perforated set Ω_ε. This geometry will need to be approximated by a sufficiently fine grid to perform the computations.

Note that the boundary of the perforated set Ω_ε is the union between the exterior Dirichlet boundary $\Gamma^D = \partial\Omega$ and the interior Neumann boundary Γ_ε^N. In this context, we take the entire boundary of the microstructures to be of Neumann type. In the construction of the geometry, the unit cell Y is supposed to have the disk inclusion Y_0 not intersecting its (exterior) boundary. We denote this exterior interface by ∂Y_{ext}, and we mean in fact $\partial Y \setminus \partial Y_0$. This is a local notation used in the code.

In the following sections, we use FEM as implemented in FEniCS to perform the needed computations.

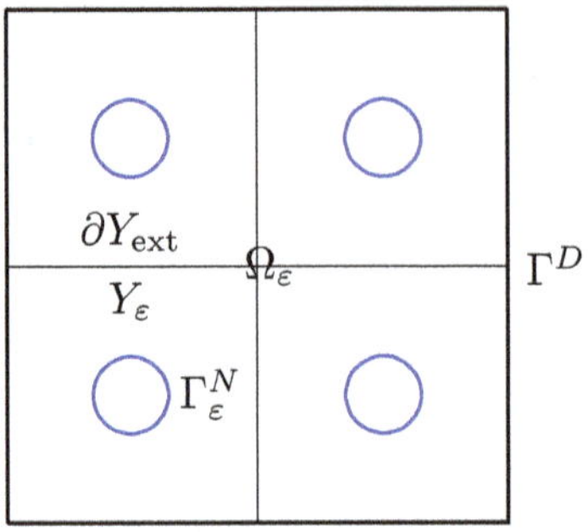

Figure 6.1. Descriptions of the reference cell Y, the macroscopic domain Ω and its perforated version, as well as the involved interfaces.

6.2.1 (P_ε) – *FEM code breakdown*

Herein, we begin by showing how to construct a mesh to handle the domain for our problem. It is important to note that the code we present here allows you to mesh for as many perforations as you want but in a certain way.

Meshing the domain for (P_ε): First, we need to load a few `Python` packages. Simply google the name you see after the import statement to find out the details regarding what each library does.

```
from dolfin import * # Read as: "from dolfin import everything", this includes Named macros which is not added otherwise
import numpy as np
import mshr
```

In the following, you can find the code that takes care of how many times we want to subdivide the domain. If we divide the domain into one part, then we simply have one cell with one hole. If we divide it twice, once horizontally and once vertically, then we will have a domain as shown in Figure 6.1. In that case, we would also have four holes, each centered in its own cell.

```
division = 1 # How many times will our domain be divided.
ep = 1/division # This will be the epsilon that we are approximating the solution for.
MeshN = 50 # A parameter that determined how many nodes there should be minimally on a (d-1)(= 1 in our case) surface.
delta = 1/6 # The radius of perforation when ep = 1
```

```
Nh = int(4**(division-1)) # This is how many holes we have in our domain as determined by the number of divisions (greater than one).
step = int(np.sqrt(Nh)) # How many holes there are in each space direction
domain = mshr.Rectangle(Point(0,0), Point(1,1)) # Base domain from which we will subtract the circles. Meshing of this domain is done later.
```

As the next step, we loop through the placements of the holes and make a domain which is the union of all the holes in our problem.

```
for i in range(step): # Loop stepping though the x-direction of our domain. i starts equal to 0 and ends as i = step-1
for j in range(step):# Loop stepping though the y-direction of our domain. j starts equal to 0 and ends as j = step-1
    if i == 0 and j == 0: # First time through the loop we need to create a circle that we add the other circles to.
        circles = mshr.Circle(Point(i/step + 0.5/step, j/step + 0.5/step), 1/step*delta) # division = 1, => Circle(Point(0.5, 0.5, 1/6))
    else: # Syntax of Circle is mshr.Circle(Point(insert x-position), Point(insert y-position), insert radius of circle)
        circles = circles + mshr.Circle(Point(i/step + 0.5/step, j/step + 0.5/step), 1/step*delta)
```

Now that we have created our domains, all we need to do is ask `dolfin` to create a mesh on the domain minus the circles. We also save the mesh as an `xdmf` file.

```
mesh = mshr.generate_mesh(domain-circles, MeshN) # Generate mesh on domain-circles
fFile = XDMFFile(MPI.comm_world, f"./Meshes/meshOriD{division}.xdmf") # Create a handle for our mesh
fFile.write(mesh) # Use that handle to write the mesh to file
```

Once this is done, the meshing of the geometry is completed. Of course, we also need meshes for the geometry of both the cell problem and the homogenized problem. The same meshing procedure needs to be done in both these cases, and hence, we omit repeating here the same information unnecessarily.

6.2.2 *Solving the microscopic problem*

The focus lies now on approximating numerically the solution to the microscopic problem (P_ε). As in the previous section, we start off by importing the libraries we need, along with performing the same construction used in the mesh script to take care of the division of the domain.

```
    import matplotlib.pyplot as plt
    import matplotlib.tri as tri
    import mshr
    import numpy as np
    import sys

    division = 1
    MacroDomainMax = 1
    num_holes = int(4**(division-1))
    ep = 1/int(np.sqrt(num_holes))
```

Next, we need to define the outer boundary Γ^D. This entails creating a function that returns "True" for any point within machine precision of the boundary.

```
    def boundary_D(x, on_boundary):
# This function defines the Dirichlet boundary. It checks if any point that fenics puts in the "on_boundary" category is on the boundary that has the Dirichlet condition. If so, it returns true
        if on_boundary:
            if near(x[0], 0) or near(x[0], MacroDomainMax) or near(x[1], 0) or near(x[1], MacroDomainMax):
                return True
            else:
                return False
        else:
            return False
```

We now load the mesh file that we created in the previous section. This involves reading the `xdmf` file.

```
# Loading the generated mesh from file
mesh = Mesh() # Create a instance of the Mesh class onto which we can read our mesh file.
f = XDMFFile(MPI.comm_world, f"./Meshes/meshOriD{division}.xdmf") # Creating the reader object
```

```
f.read(mesh) # Use the reader to connect the file to our instance
    ↪ of the Mesh class.
```

The focus now shifts to using FEniCS to describe the setting of the equations, which our mathematical model relies on. FEniCS needs the model equation in weak form.

```
U = FunctionSpace(mesh, 'P', 1) # Defining our function space --
    ↪ Lagrangean elements of first order
u = Function(U) # Define a function u in function space U
u_D = Constant(0.0)
bcD = DirichletBC(U, u_D, boundary_D) # Assemples the Diriclet
    ↪ boundary conditions by using our function "boundary_D"
```

It is now a good moment to define the model parameters, i.e. to make concrete choices for the matrix a_ε and the function f_ε.

```
a_e1 = Expression(f'2 +
    ↪ sin(x[0]*2*pi/{ep})*sin(x[1]*2*pi/{ep})', degree=2)
a_e2 = Expression(f'2 +
    ↪ sin(x[0]*2*pi/{ep})*cos(x[1]*2*pi/{ep})', degree=2)
a_e = as_matrix([[a_e1, Constant(0.0)],[Constant(0.0), a_e2]])
f = Expression(f'sin(2*pi*x[0]) +
    ↪ {ep}*sin(2*pi*x[0]/{ep})*cos(2*pi*x[1]/{ep})', degree=2)
```

At this point, we have everything we need to define explicitly our weak form and ask FEniCS to solve it for us.

```
u = TrialFunction(U) # Placeholder for the function we want to
    ↪ solve for, this is needed so that FEniCS can identify
    ↪ which function is the unknown.
v = TestFunction(U) # Test function from the same function space
a = dot(a_e*grad(u), grad(v))*dx + ep*dot(u, v)*ds +
    ↪ np.sqrt(ep)*dot(u, v)*dx # This defines our bilinear
    ↪ form.
L = f*v*dx # Righthand side
u = Function(U) # Declaring the name of our sought for function.
solve(a==L, u, bcD) # Solving a==L, with u, using Dirichlet
    ↪ boundary conditions
```

Finally, all that is left to do is to extract the vertex values and save our solution.

```python
# Extracting the vertex values of our solution and saving it as numpy array
u_vertex = u.compute_vertex_values(mesh)
np.save(f"./Solutions/OriSolD{division}", u_vertex)

# Extracts the triangulation points from the mesh (This is done mainly for the easy of plotting the solution)
n = mesh.num_vertices()
d = mesh.geometry().dim()
mesh_coordinates = mesh.coordinates().reshape((n, d))
triangles = np.asarray([cell.entities(0) for cell in cells(mesh)])
triangulation = tri.Triangulation(mesh_coordinates[:, 0],
                                  mesh_coordinates[:, 1], triangles)

# Extracts the value of u in all points of the mesh
z = np.asarray([u(point) for point in mesh_coordinates])
cmap = plt.cm.jet
plt.figure()
plt.tripcolor(triangulation, z, cmap=cmap, shading='gouraud')
plt.colorbar()
plt.savefig(f"./Figures/Solutions/OriSolD{division}.png")
```

If we do this for different values of "division" in the script by extending the discussion for different values of ε, then we can produce the following images, as shown in Figure 6.2.

6.3 Formulation of the Macroscopic Problem

We present the structure of the macroscopic (homogenized) problem (P_0), which corresponds to the periodic homogenization limit of the microscopic problem defined in Section 6.2. The plan is to use FEniCS to approximate numerically the problem (P_0), namely:

Find $u_0 \in H_0^1(\Omega)$ such that the following equations are satisfied:

$$\nabla_x\left(-\mathbb{D}\nabla_x u_0\right) + \frac{|\Gamma^N|}{|Y|}u_0 = f_0 \quad \text{in } \Omega \tag{6.4}$$

$$u_0 = 0 \quad \text{on } \Gamma^D \tag{6.5}$$

$$\mathbb{D} = \frac{1}{|Y|}\int_Y A(y)(\mathbb{I} + \nabla_y W(y))dy, \tag{6.6}$$

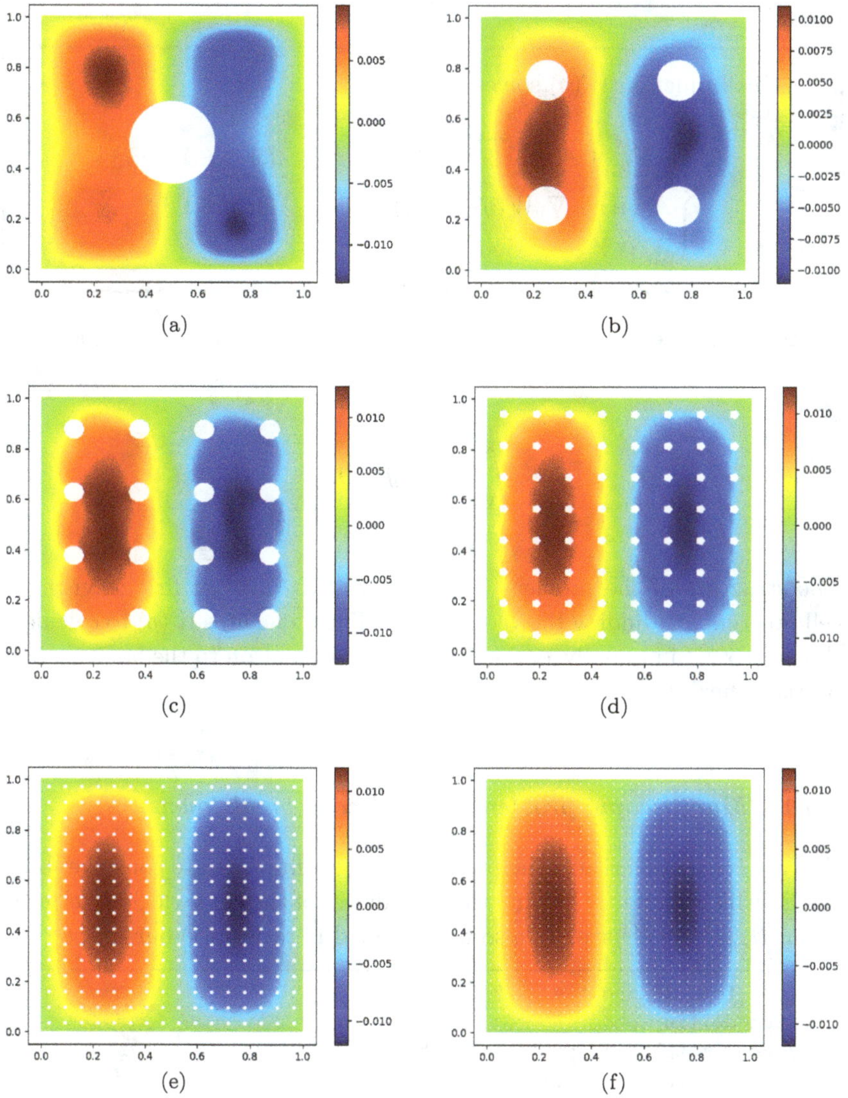

Figure 6.2. Approximate solutions of (P_ε). The images (a)–(f) are for $\varepsilon = \{1, \frac{1}{2}, \frac{1}{3}, \frac{1}{4}, \frac{1}{5}$, and $\frac{1}{6}\}$, respectively.

where

the cell function $W(y)$ solves

$$\begin{cases} -\nabla_y \cdot \Big(A(y)(\mathbb{I} + \nabla_y W(y)) \Big) = 0 & \text{for } y \in Y \\ A(y)(\mathbb{I} + \nabla_y W(y)) \cdot n(y) = 0 & \text{for } y \in \Gamma^N \cdot \\ W \text{ is } Y- \text{ periodic.} \end{cases}$$

To ensure the uniqueness property for $W(\cdot)$, we additionally impose the constraint

$$\int_Y W(y)dy = 0.$$

The structure of the problem (P_0) is typical. It involves first solving the cell problem and finding the vector $W(y)$. Then, one uses $W(y)$ to calculate the matrix $\mathbb{D}$, and finally, one solves for u_0 the solution to the homogenized partial differential equation.

6.3.1 *Macroscopic problem – FEM code breakdown*

When solving numerically the problem (P_0), we first need to construct both the Dirichlet boundary Γ^D and the boundaries of the cells which touch each other, ∂Y_{ext}. As always, we also need to load libraries into our script.

```
from fenics import *
from dolfin import *
import matplotlib.pyplot as plt
import matplotlib.tri as tri
import mshr
import numpy as np
tol = 1E-14
MacroDomainMax = 1 # Domain x_max and y_max
```

Next, we define the functions that check to which boundary a selected point belongs. We begin with those surfaces which require the implementation of periodic boundary conditions.

```
    class PeriodicBoundary(SubDomain):
# A class in FEniCS that takes "SubDomain" as an argument is expected to have one function called "inside". "inside" checks all points in "on_boundary" to see if they are on the bottom or left boundary but excludes the corner point "topleft" and "bottomright" since these point should be affected by our periodic boundary conditions. If this function returns false then "map" is called.
    # Left and bottom boundary is "target domain",
    def inside(self, x, on_boundary):
        return bool((near(x[0], 0) or near(x[1], 0)) and
                    (not ( (near(x[0], 0) and near(x[1], 1)) or
                        (near(x[0], 1) and near(x[1], 0)))) and on_boundary)
# Map simply "moves" the values that are not "inside" to be "inside".
    # Map right and top boundary (H) to left and bottom boundary (G)
    def map(self, x, y):
        if near(x[0], 1) and near(x[1], 1):
            y[0] = x[0] - 1.0
            y[1] = x[1] - 1.0
        elif near(x[0], 1):
            y[0] = x[0] - 1.0
            y[1] = x[1]
        else:
            y[0] = x[0]
            y[1] = x[1] - 1.0
```

What we have done in the code above is the most complicated thing that we need to do in order to solve our problem. Now, we only need to define our data for the cell problem. Note that the components of the elliptic matrix A are different since the homogenization step has been performed.

To have a unique solution to the cell problem, we need to impose a constraint on the vector $W(y)$. We choose the mean value of each component to be zero. We rely on the use of Lagrangian multipliers to impose this constraint. We refer the reader to Section 5 in Nepal *et al.* (2024) for a concrete example involving the Lagrange multipliers in this context. Furthermore, we prefer to solve this problem component-wise, simply because it is easier that way rather than solving directly for a vector of functions.

```
    a_11 = Expression('2+sin(2*pi*x[0])*sin(2*pi*x[1])', degree=2)
    a_22 = Expression('2+sin(2*pi*x[0])*cos(2*pi*x[1])', degree=2)
    A = as_matrix([[a_11, Constant(0.0)], [Constant(0.0), a_22]])
```

Now, we load the mesh for the cell problem and define the function spaces and functions we need.

```
mesh_w = Mesh()
f = XDMFFile(MPI.comm_world, f"./Meshes/meshCellD1.xdmf")
f.read(mesh_w)

L = FiniteElement('Lagrange', mesh_w.ufl_cell(), 1) # Lagrangian elements defined on each cell of the mesh
R = FiniteElement('Real', mesh_w.ufl_cell(), 0) # Lagrangian multiplier, real number
pbc = PeriodicBoundary() # Initializing the periodic boundary conditions
W = FunctionSpace(mesh_w, L*R, constrained_domain=pbc) # Define our function space as a product between L and R constrained by our periodic boundary conditions.
(w, c) = TrialFunction(W) # Trial and testfunctions now have two parts, Langragian element and a real value
(v, d) = TestFunction(W)
```

We define base vectors to split our problem and deal with each of them component-wise.

```
# Unit vectors to solve the cell problem component wise
e1 = as_vector([1, 0])
e2 = as_vector([0, 1])

# Weakform, cellproblem:
a_w = dot(A*grad(w), grad(v))*dx + c*v*dx + d*w*dx # Bilinear form containing the Lagrange multipliers.
L_w1 = -dot(A*e1, grad(v))*dx # Right hand side for the first component
L_w2 = -dot(A*e2, grad(v))*dx # Right hand side for the second component
```

What is left is asking FEniCS to solve our problems and extract the functions $W_1(y)$ and $W_2(y)$, which are the components of the cell function $W(y)$.

```
# Solution step, cell problem, First component of w:
W1 = Function(W)
solve(a_w==L_w1, W1) # Solving the problem for the first component of w(y)
```

```
    (w1, c) = W1.split() # Extracting w1 and discarding the constant
        ↪ c

    # Solution step, cell problem, Second component of w:
    W2 = Function(W)
    solve(a_w==L_w2, W2) # Solving the problem for the second
        ↪ component of w(y)
    (w2, c) = W2.split() # Extracting w2 and discarding the constant
        ↪ c
```

The cell problem is now solved. As the next step, we need to calculate the entries of the matrix $\mathbb{D}$. Essentially, this task amounts to evaluating numerically an integral. We do so component-wise, and finally, we obtain the explicit entries of the wanted matrix $\mathbb{D}$.

```
    # Define area of the domain
area = assemble(Constant(1.0)*dx(mesh_w)) # assemble is a numerical
    ↪ integrator.

# Computing homogenized matrix, components
a_11 = 1/area*(assemble(dot(A[0][0],
    ↪ (1+w1.dx(0)))*dx(domain=mesh_w))) # The notation "w1.dx(0)"
    ↪ means that we differentiate w1 with respect to it's first
    ↪ variable
a_12 = 1/area*(assemble(dot(A[0][0], w2.dx(0))*dx(domain=mesh_w)))
a_21 = 1/area*(assemble(dot(A[1][1], w1.dx(1))*dx(domain=mesh_w)))#
    ↪ The notation "w1.dx(1)" means that we differentiate w1 with
    ↪ respect to it's second variable
a_22 = 1/area*(assemble(dot(A[1][1], (1 +
    ↪ w2.dx(1)))*dx(domain=mesh_w)))
A_eff = as_matrix([[a_11, a_12],[a_21, a_22]]) # This is now the
    ↪ homogenized version of A.
```

6.3.2 *Solving the macroscopic problem*

The focus lies now on approximating numerically the solution to the problem (P_0). We load the mesh, define the function space, define the Dirichlet condition, and write down the weak form. We solve the problem in the same way as before. Note that the right-hand side of the target partial differential equation is now different since the second part of the right-hand side of (6.1) tends to zero in the homogenization limit.

```
    # Reading the mesh
    mesh = Mesh()
    f = XDMFFile(MPI.comm_world, f"./Meshes/meshHomo.xdmf")
    f.read(mesh)
    # Constructing the function space, functions and defining f0
    U = FunctionSpace(mesh, 'P', 1) # Lagrangian elements of order 1
    u = TrialFunction(U)
    v = TestFunction(U)
    f_0 = Expression('sin(2*pi*x[0])', degree=2)

    # boundary conditions, Dirichlet:
    u_D = Constant(0.0)
    bc = DirichletBC(U, u_D, boundary_D)

    # Weakform, homogenized problem:
    a = dot(A_eff*grad(u), grad(v))*dx(domain=mesh) +
        ↪ 2*pi*(1/6)/area*dot(u, v)*dx(domain=mesh)
    L = f_0*v*dx(domain=mesh)
```

Solve the problem, and save the results.

```
    # Solving for u_0, macro-problem.
u0 = Function(U)
solve(a==L, u0, bc)

u_vertex = u0.compute_vertex_values(mesh)
np.save(f"./Solutions/HomSol", u_vertex)

n = mesh.num_vertices()
d = mesh.geometry().dim()
mesh_coordinates = mesh.coordinates().reshape((n, d))
triangles = np.asarray([cell.entities(0) for cell in cells(mesh)])
triangulation = tri.Triangulation(mesh_coordinates[:, 0],
                                  mesh_coordinates[:, 1], triangles)

z = np.asarray([u0(point) for point in mesh_coordinates])
cmap = plt.cm.jet
plt.figure()
plt.tripcolor(triangulation, z, cmap=cmap, shading='gouraud')
plt.colorbar()
plt.savefig(f"./Figures/Solutions/HomSol.png")
```

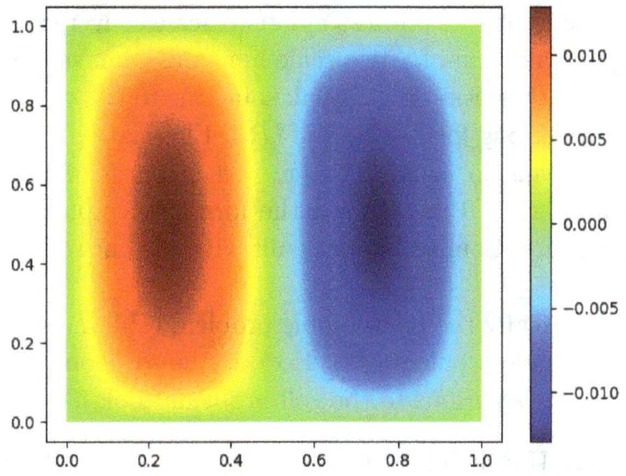

Figure 6.3. Solution of the homogenized problem (P_0).

The solution to the homogenized problem with these data is shown in Figure 6.3. Interestingly, this output seems to be very similar to the solution of the original problem when one chooses $\varepsilon = \frac{1}{6}$.

6.4 Exercises

Exercise 6.4.1. We consider the following mixed Dirichlet–Robin problem. Let Ω be a bounded domain in $\mathbb{R}^d$ with Lipschitz $\partial\Omega$. Take the perforated domain $\Omega_\varepsilon \subset \Omega$ to be connected such that the perforations with boundaries Γ_ε do not touch each other. Assume additionally that Γ_ε is Lipschitz and has $n_\varepsilon(x) = n(\frac{x}{\varepsilon})$ as the corresponding normal vector. Consider the following boundary-value problem, say (P_ε), given by

$$\begin{aligned} \operatorname{div}\big(-a_\varepsilon(x)\nabla u_\varepsilon(x)\big) &= -\sqrt{\varepsilon}u_\varepsilon(x) + f_\varepsilon(x) \quad \text{for } x \in \Omega_\varepsilon, \\ -a_\varepsilon(x)\nabla u_\varepsilon(x)\cdot n_\varepsilon(x) &= \varepsilon u_\varepsilon(x) \quad \text{for } x \in \Gamma_\varepsilon, \\ u_\varepsilon(x) &= 0 \quad \text{for } x \in \partial\Omega. \end{aligned}$$

Assume f_ε to be bounded uniformly with respect to ε in $L^2(\Omega_\varepsilon)$, and let a_ε satisfy the assumptions (H1)–(H4), as stated in Section 6.2.

(i) Show the existence and uniqueness of a weak solution u_ε to (P_ε).
(ii) Prove that u_ε is uniformly bounded in $H^1(\Omega_\varepsilon; \partial\Omega)$.
(iii) Show that there exists at least an extension of u_ε that is bounded in $H^1(\Omega)$.

(iv) Pass to the limit $\epsilon \to 0$ in (P_ε) via two-scale convergence, and determine the weak form of a two-scale limit problem, say (P_0).
(v) Prove the uniqueness of the weak solutions to (P_0).
(vi) Obtain the strong formulation of (P_0). Eliminate u_1.
(vii) Use the expansion $u_\varepsilon(x) = u_0(x, \frac{x}{\varepsilon}) + \varepsilon u_1(x, \frac{x}{\varepsilon}) + \mathcal{O}(\varepsilon^2)$ and pass to the limit $\varepsilon \to 0$ in (P_ε) via the formal asymptotic homogenization procedure. Compare your result with the answer to the previous point.
(viii) Solve numerically the microscopic problem (P_ε) and the corresponding macroscopic problem (P_0). How do the solutions compare? Illustrate how the solutions to the cell problems look like.

Exercise 6.4.2. Use FEniCS to solve the problem (P_ε) for different choices of parameters.

Exercise 6.4.3. Compute the convergence rate for the homogenization process behind the derivation of the problem (P_0) treated in this chapter. Does your numerical result agree with the theoretical estimates mentioned in Section 4.6?

Exercise 6.4.4. Extend the code presented in this chapter to be able to deal with a parabolic equation. For instance, consider the non-stationary version of the problem (P_ε), and provide non-oscillatory initial data as well.

Exercise 6.4.5. Compute the convergence rate for the homogenization process behind the derivation of the non-stationary version of the problem (P_0) treated in this chapter.

Exercise 6.4.6. Use FreeFEM++, or something else other than FEniCS, to solve the problem (P_ε) for different choices of parameters.

Exercise 6.4.7. Use the Rayleigh–Ritz–Galerkin method to approximate numerically the weak solution to (P_0) (the homogenized version of the problem (P_ε) posed in this chapter).

6.5 Solutions

Solution 6.4.1. This exercise invites the reader to perform both rigorously and formally the homogenization asymptotics for the model problem discussed in this chapter. We begin by presenting the rigorous derivation of the homogenized model using the concept of two-scale convergence adapted to perforated media (as discussed in Chapter 4). Next, we rediscover the same limiting objects using formal two-scale homogenization expansions (as outlined in Section 2.3).

(i) We show the existence and uniqueness of a weak solution to our problem by applying directly the Lax–Milgram lemma. For the convenience of writing, we introduce the notation

$$V := \{\varphi \in H^1(\Omega_\varepsilon) : \varphi = 0 \text{ at } \partial\Omega\} = H^1(\Omega_\varepsilon; \partial\Omega).$$

Let $\varphi \in V$ be an arbitrary test function. By multiplying the original microscopic problem with φ and after performing the integration by parts, the weak form of problem (P_ε) is given by the following:

Find $u_\varepsilon \in V$ such that it holds

$$\int_{\Omega_\varepsilon} a_\varepsilon \nabla u_\varepsilon \cdot \nabla\varphi \, dx + \varepsilon \int_{\Gamma_\varepsilon} u_\varepsilon \varphi \, d\sigma_\varepsilon + \sqrt{\varepsilon} \int_{\Omega_\varepsilon} u_\varepsilon \varphi \, dx = \int_{\Omega_\varepsilon} f_\varepsilon \varphi \, dx \qquad (6.7)$$

for all $\varphi \in V$. This expression can be cast into the form $\mathcal{A}(u_\varepsilon, \varphi) = \mathcal{F}(\varphi)$, where $\mathcal{A}(u_\varepsilon, \varphi)$ is the left-hand side of (6.7), while $\mathcal{F}(\varphi)$ is the right-hand side of (6.7).

Now, we investigate to which extent the hypothesis of the Lax–Milgram lemma can be verified. To do so, we use suitable Poincaré- and trace-type inequalities adapted to perforated media:

- The linear functional $\mathcal{F}(\cdot)$ is bounded. This follows directly from the fact that $f_\varepsilon \in L^2(\Omega_\varepsilon)$ and $\varphi \in V \hookrightarrow L^2(\Omega_\varepsilon)$.
- The bilinear form $\mathcal{A}(\cdot,\cdot)$ is bounded. This is guaranteed as a consequence of the estimate

$$\begin{aligned}
|\mathcal{A}(u_\varepsilon, \varphi)| &\le \int_{\Omega_\varepsilon} |a_\varepsilon||\nabla u_\varepsilon||\nabla\varphi| \, dx + \varepsilon \int_{\Gamma_\varepsilon} |u_\varepsilon||\varphi| \, d\sigma_\varepsilon + \sqrt{\varepsilon} \int_{\Omega_\varepsilon} |u_\varepsilon||\varphi| \, dx \\
&\le \beta \|\nabla u_\varepsilon\|_{L^2(\Omega_\varepsilon)} \|\nabla\varphi\|_{L^2(\Omega_\varepsilon)} + \varepsilon c_p \|\nabla u_\varepsilon\|_{L^2(\Omega_\varepsilon)} \|\nabla\varphi\|_{L^2(\Omega_\varepsilon)} \\
&\quad + \sqrt{\varepsilon} \|u_\varepsilon\|_{L^2(\Omega_\varepsilon)} \|\varphi\|_{L^2(\Omega_\varepsilon)} \\
&\le \max\left\{\beta + \varepsilon\hat{c}, \sqrt{\varepsilon} c_P\right\} \|u_\varepsilon\|_{H^1(\Omega_\varepsilon)} \|\varphi\|_{H^1(\Omega_\varepsilon)},
\end{aligned}$$

  where the constant $\hat{c}$ comes from the trace inequality for oscillating surfaces, while c_P is the Poincaré constant.
- The bilinear form $\mathcal{A}(\cdot,\cdot)$ is coercive.
  This property is proven via the following calculation. By testing the weak form with $\varphi = u_\varepsilon$, we obtain

$$\begin{aligned}
\mathcal{A}(u_\varepsilon, u_\varepsilon) &= \int_{\Omega_\varepsilon} a_\varepsilon |\nabla u_\varepsilon|^2 \, dx + \varepsilon \int_{\Gamma_\varepsilon} u_\varepsilon^2 \, d\sigma_\varepsilon + \sqrt{\varepsilon} \int_{\Omega_\varepsilon} u_\varepsilon^2 \, dx \\
&\ge \alpha \|\nabla u_\varepsilon\|^2_{L^2(\Omega_\varepsilon)} \ge \alpha \|u_\varepsilon\|^2_{H^1(\Omega_\varepsilon)}.
\end{aligned}$$

Since the hypothesis of the Lax–Milgram lemma is satisfied, we are now sure about the existence and uniqueness of the wanted weak solution.

(ii) We compute explicitly the needed *a priori* ε-independent estimates. We take $\varphi := u_\varepsilon$ as a test function in the proposed weak form. This leads to

$$\begin{aligned}\alpha\|\nabla u_\varepsilon\|^2_{L^2(\Omega_\varepsilon)} &\le \int_{\Omega_\varepsilon} a_\varepsilon(x)|\nabla u_\varepsilon(x)|^2\,dx\\ &= -\varepsilon\int_{\Gamma_\varepsilon} u_\varepsilon^2(x)\,d\sigma_\varepsilon - \sqrt{\varepsilon}\int_{\Omega_\varepsilon} u_\varepsilon^2(x)\,dx + \int_{\Omega_\varepsilon} f_\varepsilon(x)u_\varepsilon(x)\,dx\\ &\le \|f_\varepsilon\|_{L^2(\Omega_\varepsilon)}\|u_\varepsilon\|_{L^2(\Omega_\varepsilon)}\\ &\le \max\left\{\frac{1}{2},\frac{1}{2c_p}\right\}\|f_\varepsilon\|_{L^2(\Omega_\varepsilon)}\|u_\varepsilon\|_{H^1(\Omega_\varepsilon)}.\end{aligned}$$

Using the embedding $H^1(\Omega_\varepsilon) \hookrightarrow L^2(\Omega_\varepsilon)$, we obtain the bound

$$\|u_\varepsilon\|_{H^1(\Omega_\varepsilon)} \le \frac{C}{\alpha},$$

where C is a constant independent of the choice of ε.

(iii) Recalling Lemma 4.1, we deduce that there exists an extension, $\tilde{u}_\varepsilon \in H_0^1(\Omega)$, of $u_\varepsilon \in H^1(\Omega_\varepsilon; \partial\Omega)$. We also extend all the data and parameters, viz. we take $\tilde{f}_\varepsilon \in L^2(\Omega)$ as the zero extension of $f_\varepsilon \in L^2(\Omega_\varepsilon)$ and $\tilde{a}_\varepsilon \in L^2(\mathbb{R}^d)$ as the zero extension of $a_\varepsilon \in L^\infty(\Omega_\varepsilon)$. Moreover, by Lemma 4.1, there exist constants c_1 and c_2 strictly positive and independent of ε so that we can benefit from the bounds

$$\|\tilde{u}_\varepsilon\|_{H_0^1(\Omega)} \le c_1\|u_\varepsilon\|_{V(\Omega_\varepsilon)} \le c_2,$$

which allows us to continue with the passage to the two-scale convergence limit. For convenience, we keep the notation u_ε when referring to the extended version $\tilde{u}_\varepsilon$.

(iv) Now, using the obtained uniform bound in the context of Theorem 7.7, we have the following results: There exist two functions, u_0 and u_1, such that as $\varepsilon \to 0$, the following convergences hold:

$$\begin{aligned}u_\varepsilon &\overset{2}{\rightharpoonup} u_0\\ \nabla u_\varepsilon &\overset{2}{\rightharpoonup} \nabla_x u_0 + \nabla_y u_1,\end{aligned}$$

where $u_0 \in L^2(\Omega \times Y)$ and $u_1 \in L^2(\Omega; H^1_\#(Y)/\mathbb{R})$.

Take $\varphi := \varphi_0(x) + \varepsilon\varphi_1\left(x, \frac{x}{\varepsilon}\right)$, with $\varphi_0 \in C_0^\infty(\Omega)$, $\varphi_1 \in C_0^\infty(\Omega; C_\#^\infty(Y))$, and use it as test functions in the weak form of our problem. This gives

$$\int_{\Omega_\varepsilon} a_\varepsilon \nabla u_\varepsilon \cdot (\nabla_x\varphi_0 + \varepsilon\nabla_y\varphi_1)\, dx + \varepsilon \int_{\Gamma_\varepsilon} u_\varepsilon(\varphi_0 + \varepsilon\varphi_1) d\sigma_\varepsilon$$
$$+ \sqrt{\varepsilon}\int_{\Omega_\varepsilon} u_\varepsilon(\varphi_0 + \varepsilon\varphi_1)dx = \int_{\Omega_\varepsilon} f_\varepsilon(\varphi_0 + \varepsilon\varphi_1)dx.$$

Passing to the limit $\varepsilon \to 0$ by employing the concept of two-scale convergence, we obtain (via subsequences) that

$$\int_{\Omega_\varepsilon} a_\varepsilon \nabla u_\varepsilon \cdot (\nabla_x\varphi_0 + \varepsilon\nabla_y\varphi_1)\, dx \to \frac{1}{|Y|}\int_\Omega\int_Y A(y)(\nabla_x u_0(x)$$
$$+ \nabla_y u_1(x,y)) \cdot (\nabla_x\varphi_0(x) + \nabla_y\varphi_1(x,y))\, dy\, dx,$$

as well as

$$\varepsilon\int_{\Gamma_\varepsilon} u_\varepsilon(\varphi_0 + \varepsilon\varphi_1)\, d\sigma_\varepsilon \to \frac{1}{|Y|}\int_\Omega\int_\Gamma u_0\varphi_0\, d\sigma\, dx$$
$$\int_{\Omega_\varepsilon} f_\varepsilon(\varphi_0 + \varepsilon\varphi_1)dx \to \int_\Omega f_0\varphi_0\, dx,$$

while $\sqrt{\varepsilon}\int_{\Omega_\varepsilon} u_\varepsilon(\varphi_0 + \varepsilon\varphi_1)dx$ converges to zero. We did assume here that $f_\varepsilon(x) := f_0(x) + \varepsilon f_1(x/\varepsilon)$, with $\varepsilon f_1(x/\varepsilon) \to 0$ as $\varepsilon \to 0$.

Consequently, we obtain the weak formulation of the homogenized problem, say (P_0), as follows.

Find the pair $(u_0, u_1) \in H_0^1(\Omega) \times H^1(\Omega; H_\#^1(Y)/\mathbb{R})$ such that for all $(\varphi_0, \varphi_1) \in C_0^\infty(\Omega) \times C_\#^\infty(\Omega; C_\#^\infty(Y))$, the following identity holds:

$$\frac{1}{|Y|}\int_\Omega\int_Y A(y)(\nabla_x u_0 + \nabla_y u_1) \cdot (\nabla_x\varphi_0 + \nabla_y\varphi_1)\, dx\, dy$$
$$+ \frac{|\Gamma^N|}{|Y|}\int_\Omega u_0\varphi_0\, dx = \int_\Omega f_0\varphi_0\, dx.$$

Thinking of u_1 as admitting the structural assumption $u_1(x,y) = W(y)\nabla_x u_0(x)$ in terms of the cell functions $W(\cdot)$, we write

$$\frac{1}{|Y|}\int_\Omega\int_Y A(y)\left(\mathbb{I} + \nabla_y W(y)\right) dy \nabla_x u_0 \cdot \nabla_x\varphi_0\, dx$$
$$+ \frac{1}{|Y|}\int_\Omega\int_Y A(y)\left(\mathbb{I} + \nabla_y W(y)\right) dy \nabla_y\varphi_1\, dy \nabla_x u_0 dx$$
$$+ \frac{|\Gamma^N|}{|Y|}\int_\Omega u_0\varphi_0\, dx = \int_\Omega f_0\varphi_0\, dx.$$

The cell function $W = (W_1, W_2)$ (with $W_j \in H^1_\#(Y)$, $j \in \{1, 2\}$) satisfies component-wise, weakly, the elliptic boundary-value problem

$$-\nabla_y \cdot (A(y)(e_j + \nabla_y W_j(y)) = 0 \quad \text{for } y \in Y$$

$$-A(y)\nabla_y W_j(y) \cdot n(y) = 0 \quad \text{for } y \in \partial Y_0$$

$$W_j(y) \text{ is } Y\text{-periodic}, \quad \int_Y W_j(y)dy = 0.$$

This is our cell problem.

(v) We introduce the Hilbert space $(\mathcal{X}, \|\cdot\|_{\mathcal{X}})$ via

$$\mathcal{X} := H^1_0(\Omega) \times H^1(\Omega; H^1(\Omega; H^1_\#(Y))/\mathbb{R})$$

$$\|(u_0, u_1)\|^2_{\mathcal{X}} := \|\nabla_x u_0\|^2_{H^1(\Omega)} + \|\nabla_y u_1\|^2_{H^1(\Omega; H^1_\#(Y)/\mathbb{R})}.$$

Using Hölder's inequality, the triangle inequality, and suitable embeddings between the involved spaces, we can see that the expression

$$\frac{1}{|Y|}\int_\Omega \int_Y A(y)(\nabla_x u_0 + \nabla_y u_1) \cdot (\nabla_x \varphi_0 + \nabla_y \varphi_1)\, dy\, dx + \frac{|\Gamma^N|}{|Y|}\int_\Omega u_0 \varphi_0\, dx$$

builds a bounded bilinear form on $\mathcal{X}^2$. Its coercivity follows as well. In fact, as the second term is positive definite, we only need to show the wanted lower bound for the first integral. Exploiting the ellipticity of the matrix A, we get

$$\begin{aligned}
&\int_\Omega \int_Y A(y)(\nabla_x u_0 + \nabla_y u_1)(\nabla_x u_0 + \nabla_y u_1)\, dxdy \\
&\quad \geq \beta \int_\Omega \int_Y \|\nabla_x u_0 + \nabla_y u_1\|^2_{\mathbb{R}^2}\, dxdy \\
&\quad = \beta \int_\Omega \int_Y \|\nabla_x u_0\|^2_{\mathbb{R}^2}\, dxdy + \beta \int_\Omega \int_Y \|\nabla_y u_1\|^2_{\mathbb{R}^2}\, dxdy \\
&\quad\quad + 2\beta \int_\Omega \nabla u_0 \cdot \left(\int_Y \nabla_y u_1\, dy\right) dx.
\end{aligned}$$

Since u_1 is Y-periodic, the last term in the above right-hand side vanishes after applying Green's formula. This calculation yields

$$\begin{aligned}
&\textstyle\int_\Omega \int_Y A(y)(\nabla_x u_0 + \nabla_y u_1)(\nabla_x u_0 + \nabla_y u_1)\, dxdy \\
&\quad \geq \beta|Y|\|\nabla u_0\|^2_{H^1(\Omega)} + \beta\|\nabla_y u_1\|^2_{H^1_\#(\Omega\times Y)} \\
&\quad \geq \min\{\beta|Y|, \beta\}\|(u_0, u_1)\|^2_{\mathcal{X}},
\end{aligned}$$

and thus the bilinear form is coercive on $\mathcal{X}^2$. The hypothesis of the Lax–Milgram lemma is satisfied, so the existence and uniqueness of a weak solution to problem (P_0) is ensured.

(vi) Setting $\varphi_0 = 0$ in the weak form (P_0), we have

$$\frac{1}{|Y|}\int_\Omega\int_Y A(y)(\mathbb{I} + \nabla_y W(y))\nabla_y\varphi_1\,dy\nabla_x u_0\,dx = 0.$$

Integrating this result by parts, we get that

$$\frac{1}{|Y|}\int_\Omega\int_Y -\nabla_y\cdot(A(y)(\mathbb{I} + \nabla_y W(y)))\,\varphi_1\,dy\nabla_x u_0\,dx$$
$$+\frac{1}{|Y|}\int_\Omega\int_{\partial Y} A(y)(\mathbb{I} + \nabla_y W(y))\cdot n(y)\varphi_1\,d\sigma_y\nabla_x u_0\,dx = 0,$$

which can now be used to discover the structure of the cell problem.

Next, take $\varphi_1 = 0$ in the weak form (P_0) and arrive at

$$-\int_\Omega \nabla_x\cdot\left[\frac{1}{|Y|}\int_Y A(y)(\mathbb{I} + \nabla_y W(y))\,dy\right] u_0\nabla_x\varphi_0\,dx + \frac{|\Gamma^N|}{|Y|}\int_\Omega u_0\varphi_0\,dx$$
$$= \int_\Omega f_0\varphi_0\,dx,$$

which gives us the macroscopic problem:

Find $u_0 \in H^1(\Omega;\partial\Omega)$ such that

$$-\nabla_x\cdot\left(\frac{1}{|Y|}\int_Y A(y)\left(\mathbb{I} + \nabla_y W(y)\right)dy\nabla_x u_0\right) + \frac{|\Gamma^N|}{|Y|}u_0 = f_0 \quad \text{in } \Omega,$$
$$u_0 = 0 \quad \text{on } \partial\Omega.$$

The effective diffusion coefficient is, as expected,

$$\mathbb{D} := \frac{1}{|Y|}\int_Y A(y)\left(\mathbb{I} + \nabla_y W(y)\right)dy.$$

(vii) We work now with the standard homogenization *Ansatz*:

$$u_\varepsilon(x) = u_0(x, \frac{x}{\varepsilon}) + \varepsilon u_1(x, \frac{x}{\varepsilon}) + \varepsilon^2 u_2(x, \frac{x}{\varepsilon}) + \text{h.o.t},$$

where the terms u_k are periodic in their second variable for all $k \in \mathbb{N}$. Substituting this *Ansatz* into the microscopic problem, we obtain

$$\left(\nabla_x + \frac{1}{\varepsilon}\nabla_y\right)\cdot\left(-A(y)\left(\frac{1}{\varepsilon}\nabla_y u_0 + \varepsilon^0(\nabla_x u_0 + \nabla_y u_1) + \varepsilon^1(\nabla_x u_1 + \nabla_y u_2)\right)\right)$$
$$= -\sqrt{\varepsilon}\left(u_0 + \varepsilon u_1 + \varepsilon^2 u_2\right) + f_0 + \varepsilon f_1 + \text{h.o.t.}$$

Collecting the terms with the same power of ε and then letting $\varepsilon \to 0$, we get the structures

$$\begin{aligned} T_{-2}(u_0) &:= \nabla_y \cdot (-A(y)\nabla_y u_0) = 0 \\ T_{-1}(u_0, u_1) &:= \nabla_y \cdot (-A(y)\nabla_x u_0 - A(y)\nabla_y u_1) + \nabla_x \cdot (-A(y)\nabla_y u_0) = 0 \\ T_0(u_0, u_1, u_2) &:= \nabla_x \cdot (-A(y)\nabla_x u_0 - A(y)\nabla_y u_1) + \nabla_y \cdot (-A(y)\nabla_x u_1 \\ &\quad - A(y)\nabla_y u_2) = f_0. \end{aligned}$$

Regarding the Dirichlet boundary condition, we impose $u_0(x, \frac{x}{\varepsilon}) + \varepsilon u_1(x, \frac{x}{\varepsilon}) + \varepsilon^2 u_2(x, \frac{x}{\varepsilon}) + \text{h.o.t} = 0$ at $\partial\Omega$. This implies $0 = u_0(x)$ for $x \in \partial\Omega$. Using the proposed *Ansatz* to handle the information at the interfaces Γ_ε leads to

$$\varepsilon u_\varepsilon = (-a_\varepsilon \nabla u_\varepsilon) \cdot n_\varepsilon = \left(-a_\varepsilon \left(\nabla_x + \frac{1}{\varepsilon}\nabla_y\right) \sum_{i=0}^{\infty} \varepsilon^i u_i(x, y)\right) \cdot n_\varepsilon.$$

Grouping the terms with the same order of ε, we note that

$$\begin{aligned} &-\frac{1}{\varepsilon^2} a_\varepsilon \nabla_y u_0 - \frac{1}{\varepsilon}(a_\varepsilon \nabla_x u_0 - a_\varepsilon \nabla_y u_1) - \varepsilon a_\varepsilon \nabla_x u_1 - \varepsilon a_\varepsilon \nabla_y u_2 + \text{h.o.t} \\ &\qquad = \varepsilon(u_0 + \varepsilon u_1 + \varepsilon^2 u_2) + \text{h.o.t.} \end{aligned}$$

This leads to the following relations across Γ, holding as ε approaches zero:

$$\begin{aligned} T_{-2}: &\quad -a(y)\nabla_y u_0(x, y) \cdot n_\varepsilon = 0, \\ T_{-1}: &\quad (-a(y)\nabla_x u_0(x, y) - a(y)\nabla_y u_1(x, y)) \cdot n_\varepsilon = 0, \\ T_0: &\quad (-a(y)\nabla_x u_1(x, y) - a(y)\nabla_y u_2(x, y)) \cdot n_\varepsilon = u_0. \end{aligned}$$

The auxiliary problem T_{-2} reads as follows:

Find $u_0 = u_0(x, y)$ such that

$$\begin{aligned} \nabla_y \cdot (-A(y)\nabla_y u_0(x, y)) = 0 &\quad \text{in } Y \\ -A(y)\nabla_y u_0(x, y) \cdot n(y) = 0 &\quad \text{at } \Gamma, \end{aligned}$$

where $u_0(x, \cdot)$ is Y-periodic. We deduce from here that $u_0 = u_0(x)$.

The auxiliary problem T_{-1} reads as follows:
Find $u_1 = u_1(x, y)$ such that

$$\nabla_y \cdot (-A(y)\nabla_x u_0 - A(y)\nabla_y u_1) + \nabla_x \cdot (-A(y)\nabla_y u_0) = 0 \quad \text{in } Y$$
$$(-A(y)\nabla_x u_0 - A(y)\nabla_y u_1) \cdot n(y) = 0 \quad \text{at } \partial Y_0,$$

where $u_1(x, \cdot)$ is Y-periodic. This problem has a unique weak solution if and only if

$$\int_Y \nabla_y \cdot A(y)\nabla_x u_0(x)dy = \int_\Gamma (A(y)\nabla_x u_0(x)) \cdot n(y)d\sigma_y.$$

The weak solution $u_1 = u_1(x, y)$ exists uniquely. Using the cell functions, we rewrite u_1 as $u_1(x, y) = W(y) \cdot \nabla_x u_0(x)$. Lastly, we consider the auxiliary problem T_0, which reads as follows:
Find $u_2 = u_2(x, y)$ such that

$$-A(y)\nabla_x u_1 - A(y)\nabla_y u_2 = f_0 \quad \text{in } Y$$
$$(-A(y)\nabla_x u_1 - A(y)\nabla_y u_2) \cdot n(y) = u_0 \quad \text{at } d\Gamma,$$

where $u_2(x, \cdot)$ is Y periodic.
The standard solvability condition leads to the following structure of the macroscopic problems:

$$-\nabla_x \cdot \left(\frac{1}{|Y|}\int_Y A(y)\left(\mathbb{I} + \nabla_y w(y)\right) dy \nabla_x u_0\right) + \frac{|\Gamma^N|}{|Y|} u_0 = f_0 \quad \text{in } \Omega$$
$$u_0 = 0 \quad \text{on } \partial\Omega$$

with

$$\mathbb{D} := \frac{1}{|Y|}\int_Y A(y)\left(\mathbb{I} + \nabla_y W(y)\right) dy.$$

This result is in perfect agreement with what we obtained using the two-scale convergence procedure.

(viii) To fix ideas, we take the next concrete choice of oscillating functions:

$$\begin{aligned} f_\varepsilon(x, x/\varepsilon) &:= f_0(x) + \varepsilon f_1(x, x/\varepsilon) \\ &= 2 + \sin(2\pi x_1) + \varepsilon \sin(2\pi x_1/\varepsilon)\cos(2\pi x_2/\varepsilon), \end{aligned}$$

$$A(x/\varepsilon) := \begin{bmatrix} 2 + \sin(2\pi x_1/\varepsilon)\sin(2\pi x_2/\varepsilon) & 0 \\ 0 & 2 + \sin(2\pi x_1/\varepsilon)\cos(2\pi x_2/\varepsilon) \end{bmatrix},$$

where $x = (x_1, x_2) \in \Omega_\varepsilon$.

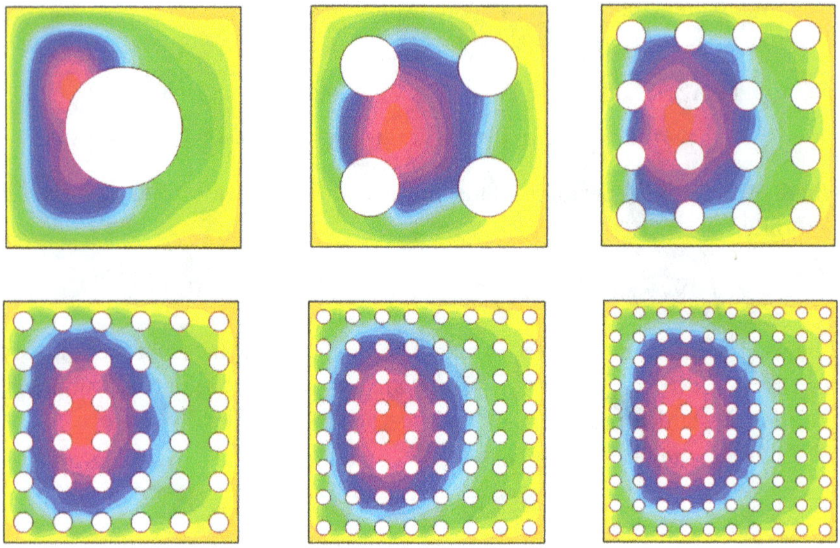

Figure 6.4. Approximate solution of (P_ε) for various choices of $\varepsilon \in \{1, \frac{1}{2}, \frac{1}{4}, \frac{1}{6}, \frac{1}{8}, \frac{1}{10}\}$.

Firstly, we solve the original microscopic problem (P_ε) for various choices of the small parameter ε. The radius of the perforation (i.e. of the inner disk) is taken to be $r(\varepsilon) = \frac{1}{4}\varepsilon$. We plot the corresponding numerical results in Figure 6.4.

Next, we focus our attention on the numerical approximation of the homogenized macroscopic problem (P_0).

The plot of the solution to the problem (P_0) is given in Figure 6.5, while the solution of the two cell problems are shown in Figure 6.6. Comparing Figures 6.4 and 6.5, we observe that smaller values for ε used when computing the original problem P_ε give results that are closer to the solution of the problem (P_0). This confirms what we know theoretically (cf. (i)–(vii)) and can be used to quantify the quality of the performed averaging.

It is also worth observing that the numerical results indicate that the boundary conditions were correctly implemented. Looking, for instance, at Figures 6.4 and 6.5, we clearly see that the homogeneous Dirichlet condition holds at $\partial\Omega$.

Figure 6.6 indicates that the periodicity property of the cell functions $W_1(\cdot)$ and $W_2(\cdot)$ is correctly implemented.

Figure 6.5. Solution of the homogenized problem (P_0). Note that the homogeneous Dirichlet condition appears to hold along $\partial\Omega$.

Figure 6.6. Solution of the cell problems $W_1(\cdot)$ (left) and, respectively, $W_2(\cdot)$ (right). Note that the periodicity of the functions is satisfied.

Solution 6.4.2. Make sure that you have `Python` and FEniCS installed. Use the codes listed in this chapter to build your own code. Modify the code according to your preferred selection of parameters.

Solution 6.4.6. See the appendix for an example of a FreeFEM++ code addressing this matter.

Solution 6.4.7. Assume that the target solution admits the form (7.3), and then solve for the coefficients c_i in (7.4) for a suitable Galerkin basis (ϕ_i); see Section 7.3 for more context about the application of the Rayleigh–Ritz–Galerkin method to elliptic equations.

Chapter 7

Toolbox in Applied Functional Analysis

We review the basic functional analytic notions[1] that the reader needs to be familiar with to be able to derive, in a rigorous manner, upscaled model equations and effective coefficients. Since the problems treated in this textbook are linear, the key concept helping to pass to the limit $\varepsilon \to 0$ is connected to some kind of weak convergence adapted to handling oscillating periodic functions. The discussion is generally done in terms of Sobolev and Bochner spaces. Furthermore, the Lax–Milgram lemma and Fredholm's alternative are important results in this context. We recall briefly these and related theoretical aspects and give hints to the specialized literature where more details can be found.

We use the standard notation for Lebesgue and Sobolev spaces. We denote these spaces in the same way for both scalar and vector cases, i.e. we use $L^p(\Omega)$ instead of $L^p(\Omega; \mathbb{R}^3)$. Furthermore, $c > 0$ is a constant independent of ε, possibly having a different value whenever it appears in text.

7.1 Spatial Domains in Homogenization Settings

Let $\emptyset \neq \Omega \subset \mathbb{R}^d$ be a bounded open set. As typical for homogenization settings, $Y := [0,1]^d$ is the unit hypercube which can be identified with the unit torus $\mathbb{R}^d/\mathbb{Z}^d$.

In the case of perforated domains, we prefer to use the notation $Z := [0,1]^d$ and $Y := Z \setminus \bar{Y}_0$, where $Y_0 \subset Z$ is an open set with smooth boundary

[1] Ideally, the reader has previously followed a course related to introductory notions of functional analysis and a beginner course in partial differential equations.

$\Gamma = \partial Y_0$ not intersecting ∂Z. We identify Y_0, Y, and Γ with their extensions by Y-periodicity to the whole $\mathbb{R}^d$ (eventually restricted to Ω). Then, as given by Allaire *et al.* (1995) and Ainouz (2007), for a decreasing-to-zero sequence of positive numbers ε, we define the perforated domain Ω_ε as

$$\Omega_\varepsilon := \left\{x \in \Omega : \frac{x}{\varepsilon} \in Y\right\}.$$

The corresponding oscillating hypersurface Γ_ε is

$$\Gamma_\varepsilon := \left\{x \in \Omega : \frac{x}{\varepsilon} \in \Gamma\right\}.$$

Clearly, it holds that $\Gamma_\varepsilon = \partial\Omega_\varepsilon$. To fix notation, we denote by $d\sigma_y$ and $d\sigma_\varepsilon$ the surface measures on Γ and Γ_ε, respectively.

7.2 Basic Function Spaces

Let B be a real Banach space equipped with the norm $\|\cdot\|_B$. Correspondingly, B^* refers to the dual space of B, namely

$$B^* := \mathcal{L}(B, \mathbb{R}) = \{\zeta : B \to \mathbb{R} \mid \zeta \text{ linear and continuous}\}.$$

We endow the space B^* with the norm

$$\|\zeta\|_{B^*} = \sup_{x \in B\setminus\{0\}} \frac{|\langle \zeta, x\rangle_{B^*,B}|}{\|x\|_B}, \tag{7.1}$$

for all $\zeta \in B^*$. In (7.1), for a $\zeta \in B^*$, the dual pairing $\langle \zeta, x\rangle_{B^*,B}$ represents the image $\zeta(x)$ of an element $x \in B$.

Let B^{**} denote the bidual space of B, i.e. $(B^*)^* = B^{**}$. The canonical map $B \ni b \mapsto b^{**} \in B^{**}$, defined by $b^{**}(b^*) = b^*(b)$ for $b^* \in B^*$, forms an isometric linear isomorphism between B and B^{**}. If the canonical map between B and B^{**} is surjective, then the Banach space B is called reflexive.

Within the scope of this textbook, the Banach space B is, for instance, one of the following examples: $C_0^\infty(\Omega)$, $L^p(\Omega)$, $H_0^1(\Omega)$, $W^{1,p}(\Omega)$, $L^2_\#(Y, C(\bar{\Omega}))$, $H^1_\#(Y)$, $L^2(\Omega,\ H^1_\#(Y))$, and $L^2(0,T;H^1(\Omega))$, where Ω and Y are open sets in $\mathbb{R}^d$ with Lipschitz boundary, with $d \in \{1,2,3\}$ and $p \in [1,\infty]$. Observe, for instance, that all the spaces $L^p(\Omega)$ with $p \in (1,+\infty)$ are reflexive Banach spaces, while the spaces $L^1(\Omega)$, $L^\infty(\Omega)$, and $C(\bar{\Omega})$ are examples of non-reflexive Banach spaces.

In this textbook, most of the discussions about weak solutions of linear elliptic partial differential equations refer to the space $H_0^1(\Omega)$. However, relatively easy to make adaptations are possible for the case of $W_0^{1,p}(\Omega)$.

Note that $L^2(\Omega)$ and $H^1(\Omega)$ (or $H_0^1(\Omega)$ and $L^2(\Omega; H_\#^1(Y))$) are all Hilbert spaces. We often use the shorthand notation H when referring to a Hilbert space. Then, $\langle \zeta, x \rangle_{H^*,H} = (\zeta, x)_H$ and $H^* = H$ (in the sense of the Riesz representation theorem). Here, $(\alpha, \beta)_H$ is the scalar product of the elements α and β taken from the Hilbert space H. To further emphasize the structure of the space, if H is a Hilbert space, then we always denote generically by $||\cdot||_H$ a norm in H and by $(\cdot, \cdot)_H$ a scalar product in H.

Note that the Hilbert spaces are examples of reflexive Banach space.

The space $H_\#^1(Y)$ plays a prominent role. If we denote by $C_\#^\infty(Y)$ the subset of $C^\infty(\mathbb{R}^d)$ of Y-periodic functions, then the space $H_\#^1(Y)$ is simply the closure of $C_\#^\infty(Y)$ taken with respect to the H^1-norm.

The quotient space

$$H_\#^1(Y)/\mathbb{R}$$

is the space of equivalence classes with respect to the relation $\mathcal{R}$, which is defined by

$$u\mathcal{R}v \text{ if and only if } u - v = c \in \mathbb{R} \quad \text{for } u, v \in H_\#^1(Y).$$

By Proposition 3.52 in Cioränescu and Donato (1999), the space $H_\#^1(Y)/\mathbb{R}$ can be endowed with the norm

$$||v||_{H_\#^1(Y)/\mathbb{R}} := ||\nabla v||_{L^2(\Omega)}.$$

Additionally, it is worth noting that some of the mentioned function spaces have a nice structure. For instance, it is useful to know the next identifications of spaces

$$\begin{aligned} L^2(\Omega \times Y) &= L^2(Y \times \Omega) = L^2(\Omega; L^2(Y)), \\ H^1(\Omega \times Y) &= H^1(Y \times \Omega) = H^1(\Omega; H^1(Y)), \\ L^2(\Omega, H_\#^1(Y) &= H_\#^1(Y, L^2(\Omega)). \end{aligned}$$

The proofs of these properties rely on suitable applications of Fubini's theorem.

The spaces of squared integrable functions with respect to the measures $d\sigma_y$ and $d\sigma_\varepsilon$ are denoted by $L^2(\Gamma)$ and $L^2(\Gamma_\varepsilon)$, respectively. If Γ is identified with its periodic extension, then $L^2(\Gamma)$ and $L_\#^2(\Gamma)$ refer to the same set of functions.

7.3 Lax–Milgram Lemma and Related Results

Theorem 7.1 (Lax–Milgram lemma). *Let H be a Hilbert space. Consider the bilinear functional $\mathcal{A} : H \times H \to \mathbb{R}$ and the linear functional $\mathcal{F} : H \to \mathbb{R}$ satisfying the conditions:*

(i) *$\mathcal{A}$ is coercive (or H-elliptic), i.e. there exists $\alpha > 0$ such that*

$$\mathcal{A}(v, v) \geq \alpha ||v||_H^2$$

for all $v \in H$;

(ii) *$\mathcal{A}$ is continuous, i.e. there exists $C > 0$ such that*

$$|\mathcal{A}(v_1, v_2)| \leq C||v_1||_H||v_2||_H$$

for all $v_1, v_2 \in H$;

(iii) *$\mathcal{F}$ is continuous, i.e. there exists $c > 0$ such that*

$$|\mathcal{F}(v)| \leq c||v||_H$$

for all $v \in H$.

Then, there exists a unique solution to the equation: Find $u \in H$ such that

$$\mathcal{A}(u, v) = \mathcal{F}(v) \quad \text{for all } v \in H$$

is solvable. Furthermore, the solution u satisfies the energy estimate

$$||u||_H \leq \frac{c}{\alpha} \tag{7.2}$$

for some $c > 0$.

If $\mathcal{A}$ is symmetric, i.e. $\mathcal{A}(v_1, v_2) = \mathcal{A}(v_2, v_1)$ for all $v_1, v_2 \in H$, then $u \in H$ is the unique minimizer of the functional $\mathcal{J} : H \to \mathbb{R}$, defined by

$$\mathcal{J}(v) := \frac{1}{2}\mathcal{A}(v, v) - \mathcal{F}(v).$$

The hypotheses present in this formulation of the Lax–Milgram lemma can be relaxed. For instance, there are useful generalizations of Theorem 7.1, particularly due to I. Babuska (allowing the use of a bilinear mapping $\mathcal{A} : H_1 \times H_2 \to \mathbb{R}$, defined now on possibly two different Hilbert spaces H_1 and H_2) and to J.-L. Lions (allowing the use of a bilinear mapping $\mathcal{A} : H \times V \to \mathbb{R}$, defined now on the cross product between a Hilbert space H and a normed space V). The assumptions on $\mathcal{F}$ simply state that $\mathcal{F} \in H^*$.

We find it quite interesting that Theorem 7.1 can also be used outside the framework of linear elliptic equations – one can use it, for instance, to handle the well-posedness of a linear second-order parabolic problem; see Showalter (1997) for an example of such an application.

The Lax–Milgram lemma is a very useful tool in the investigation of boundary-value problems of type

$$Au(x) = f(x), \quad x \in \Omega;$$
$$Bu(x) = g(x), \quad x \in \partial\Omega,$$

where A and B are suitable linear operators. It is used repeatedly throughout the textbook. Interestingly, besides ensuring the well-posedness of such boundary-value problems, it also helps in the application of a computational tool – the Rayleigh–Ritz–Galerkin method. This method deals with finding an approximate solution u_M of $u \in H$, solving the weak form $\mathcal{A}(u, v) = \mathcal{F}(v)$ for all $v \in H$ (or of the associated operator equation) in the form of a finite series

$$u_M = \sum_{i=1}^{M} c_i\phi_i + \phi_0 \quad \text{for } i \in \{1, \ldots, M\}. \tag{7.3}$$

The coefficients c_i arising in (7.3) are called the Rayleigh–Ritz–Galerkin coefficients. They are chosen such that the weak form $\mathcal{A}(u, v) = \mathcal{F}(v)$ holds for all $v = \phi_i$ for $i \in \{1, \ldots, M\}$, viz.

$$\mathcal{A}\left(\sum_{i=1}^{M} c_i\phi_i + \phi_0, \phi_i\right) = \mathcal{F}(\phi_i) \quad \text{for } i \in \{1, \ldots, M\}. \tag{7.4}$$

The elements ϕ_i are given; see Section 22.1 in the work of Zeidler (1990) for remarks on how to choose them.

The reader might find it useful to refresh the knowledge about Fredholm's alternative formulated for elliptic boundary-value problems with homogeneous Dirichlet boundary conditions (cf. e.g. Evans (2015, Theorem 4 p. 323)) and those with periodic boundary conditions (cf. e.g. Pavliotis and Stuart (2008, pp. 108–113)). Fredholm alternative-type results are very useful tools, particularly when handling auxiliary problems that arise in the application of two-scale formal asymptotic expansions; see Section 2.3.

We briefly point out how Fredholm's alternative applies to matrices and formally translate the statements for linear operators. In this spirit, the equation $Au = b$ has a solution if and only if $(b, v) = 0$ for all vectors v satisfying $A^*v = 0$. If the vector u exists such that the equation $Au = b$ is satisfied, then this solution is also unique if and only if $u = 0$ is the only solution of $Au = 0$. If instead of being a matrix, A is now a bounded operator on a Hilbert space H, then the operator equation $Au = f$ has a solution if and only if $(f, v)_H = 0$ for all $v \in H$ that also fulfill $A^*v = 0$. The statement translates in a direct manner to boundary-value problems, but it eventually involves boundary terms as

$$(Au, v)_H = \mathcal{B}(u, v) + (u, A^*v)_H,$$

where the term $\mathcal{B}(u, v)$ marks all involved boundary terms. With this notation at hand, the boundary-value problem

$$\begin{aligned} Au(x) &= f(x), \quad x \in \Omega, \\ Bu(x) &= g(x), \quad x \in \partial\Omega, \end{aligned}$$

(with A and B suitable linear operators) admits a solution if and only if

$$(f, v)_H = \mathcal{B}(u, v)$$

for all $v \in H$ which fulfill $A^*v = 0$ and $B^*v = 0$.

The following corollary of Fredholm's alternative (cf. e.g. Persson *et al.* (1993, Lemma 2.1)) is often used when performing formal two-scale asymptotic homogenization expansions for elliptic and parabolic second-order partial differential equations.

Lemma 7.1. *Let $F \in L^2(Y)$ and $g \in L^2(\partial Y)$. Consider the following problem: Find $\omega \in H^1_\#(Y)$ satisfying weakly the boundary-value problem*

$$\nabla \cdot \big(-A(y)\nabla\omega\big) = F(y) \quad \text{for } y \in Y, \tag{7.5}$$

$$-A(y)\nabla\omega \cdot n = g(y) \quad \text{for } y \in \partial Y, \tag{7.6}$$

$$\omega \text{ is } Y-\text{periodic.} \tag{7.7}$$

Then, the following statements hold:

(i) *There exists a weak Y-periodic solution $\omega \in H^1_\#(Y)$ to (7.5) if and only if $\int_Y F(y)dy = \int_{\partial Y} g(y)d\sigma_y$.*
(ii) *If the first statement holds, then the uniqueness of weak solutions is ensured up to an additive constant.*

7.4 Weak and Strong Convergences

We introduce a generic sequence of positive numbers $(\varepsilon_n) \subset \mathbb{R}_+$ such that $\lim_{n\to\infty} \varepsilon_n = 0$. We abbreviate (ε_n), together with all its convergent subsequences, by calling it simply ε.

Definition 7.1 (Weak convergence). A sequence $(u_\varepsilon) \subset B$ is said to converge weakly to $u \in B$ as $\varepsilon \to 0$ if and only if

$$\text{for all } \zeta \in B^* \text{ we have } \langle \zeta, u_\varepsilon \rangle_{B^*,B} \to \langle \zeta, u \rangle_{B^*,B} \quad \text{as } \varepsilon \to 0. \tag{7.8}$$

We denote the convergence (7.8) by $u_\varepsilon \rightharpoonup u$ and mean implicitly that $\varepsilon \to 0$.

Let $p \in (1, \infty)$. For the choice of $B = L^p(\Omega) \ni u_\varepsilon$, the expression $u_\varepsilon \rightharpoonup u$ in $L^p(\Omega)$ means that

$$\int_\Omega u_\varepsilon \varphi dx \to \int_\Omega u \varphi dx$$

for all $\varphi \in L^q(\Omega) = (L^p(\Omega))^*$, with $\frac{1}{p} + \frac{1}{q} = 1$.

If, on the other hand, we consider the case $p = \infty$, then for the choice of $B = L^\infty(\Omega) \ni u_\varepsilon$, the expression $u_\varepsilon \overset{*}{\rightharpoonup} u$ means that

$$\int_\Omega u_\varepsilon \varphi dx \to \int_\Omega u \varphi dx$$

for all $\varphi \in L^1(\Omega)$. This is what we refer to as weak* convergence.

Note that since $L^1(\Omega)$ is a non-reflexive Banach space, the concepts of weak convergence and weak* convergence in $L^\infty(\Omega)$ are not equivalent.

Theorem 7.2 (Strong convergence). *Let B be a uniformly convex Banach space and take $(u_\varepsilon) \subset B$. We have $u_\varepsilon \to u$ (strongly) in B if and only if:*

(i) $u_\varepsilon \rightharpoonup u$ *in* B;
(ii) $\|u_\varepsilon\|_B \to \|u\|_B$.

Standard examples of Banach spaces that are uniformly convex are $L^p(\Omega)$ for any $p \in (1, \infty)$. The space $L^\infty(\Omega)$ is not uniformly convex. For more information on the concepts of weak and strong convergence and their properties, see for instance Ciorănescu and Donato (1999) and Evans (2015).

The remaining results mentioned in this section point out specific features for convergent sequences of periodic functions.

Lemma 7.2. *Let $a \in L^\infty(\mathbb{R})$, a 1-periodic, and define $a_\varepsilon(x) = a\left(\frac{x}{\varepsilon}\right)$ for $x \in \mathbb{R}$. Then, $a_\varepsilon \overset{*}{\rightharpoonup} \langle a \rangle$ in $L^\infty(\mathbb{R})$ as $\varepsilon \to 0$.*

Proof. We choose to give a direct proof for this statement.[2]

Recall that $a_\varepsilon \overset{*}{\rightharpoonup} \langle a\rangle$ in $L^\infty(\mathbb{R})$ as $\varepsilon \to 0$ means that for all $\varphi \in L^1(\mathbb{R})$, it holds that

$$\int_{\mathbb{R}} a_\varepsilon(x)\varphi(x)dx \to \int_0^1 a(x)dx \int_{\mathbb{R}} \varphi(x)dx$$

as $\varepsilon \to 0$. Observe that, instead of working with any test function $\varphi \in L^1(\mathbb{R})$, we can take the characteristic function of any set $[\alpha,\beta] \subset [0,1]$ with $\alpha < \beta$, i.e.

$$\varphi(x) = \chi_{[\alpha,\beta]}(x) = \begin{cases} 1, & x \in [\alpha,\beta], \\ 0, & x \in \mathbb{R}\backslash[\alpha,\beta]. \end{cases}$$

Then, as the next step, we can argue relying on the density of step functions in $L^1(\mathbb{R})$. So, keeping in mind that for all $x \in \mathbb{R}$, it holds that[3] $x = [x]+\{x\}$, we have the following identity:

$$\begin{aligned}\int_{\mathbb{R}} a_\varepsilon(x)\varphi(x)dx &= \int_{\mathbb{R}} a\left(\frac{x}{\varepsilon}\right)\chi_{[\alpha,\beta]}(x)dx = \int_\alpha^\beta a\left(\frac{x}{\varepsilon}\right)dx \\ &= \varepsilon\int_{\frac{\alpha}{\varepsilon}}^{\frac{\beta}{\varepsilon}} a(y)dy = \varepsilon\left[\int_0^{\frac{\beta}{\varepsilon}} a(y)dy - \int_0^{\frac{\alpha}{\varepsilon}} a(y)dy\right] \\ &= \varepsilon\left(\left[\frac{\beta}{\varepsilon}\right]\langle a\rangle + r_\varepsilon(\beta)\right) - \varepsilon\left(\left[\frac{\alpha}{\varepsilon}\right]\langle a\rangle + r_\varepsilon(\alpha)\right) \\ &= \varepsilon\left(\left[\frac{\beta}{\varepsilon}\right] - \left[\frac{\alpha}{\varepsilon}\right]\right)\langle a\rangle + \varepsilon\left(r_\varepsilon(\beta) - r_\varepsilon(\alpha)\right).\end{aligned}$$

Take $x \in (0,1)$. It is now important to note the following ingredients, which are all behind the structure of the previous identity:

(1) Observe that $\varepsilon\int_0^{\frac{x}{\varepsilon}} a(s)ds = \varepsilon\left([\frac{x}{\varepsilon}]\langle a\rangle + r_\varepsilon(x)\right)$, where

$$r_\varepsilon(x) := \int_{[\frac{x}{\varepsilon}]}^{\frac{x}{\varepsilon}} a(s)ds.$$

[2]This is a basic result in linear functional analysis. We prove it here because of a twofold reason: on the one hand, this convergence result for periodic functions is very much linked to the core ideas of any course in averaging techniques; on the other hand, the proof involves scaling arguments typical to an asymptotic study. A careful reading of this proof could be planned as a classroom exercise.

[3]Note also the cute fact that the function $f : \mathbb{R} \to \mathbb{R}$, defined by $f(x) = x - [x]$, is 1-periodic.

(2) For all $x \in (0,1)$, it holds that $|r_\varepsilon(x)| \leq ||a||_{L^\infty(\mathbb{R})}$.
(3) Since

$$\frac{x}{\varepsilon} - 1 \leq \left[\frac{x}{\varepsilon}\right] \leq \frac{x}{\varepsilon} + 1,$$

we deduce that $\varepsilon[\frac{x}{\varepsilon}] \to x$ as $\varepsilon \to 0$.

Putting together the facts yields

$$\langle a \rangle \varepsilon \left(\frac{\beta}{\varepsilon} - 1 - \frac{\alpha}{\varepsilon}\right) \leq -\varepsilon \left(r_\varepsilon(\beta) - r_\varepsilon(\alpha)\right) + \varepsilon \int_{\frac{\alpha}{\varepsilon}}^{\frac{\beta}{\varepsilon}} a(y)dy$$
$$\leq \varepsilon \left(\frac{\beta}{\varepsilon} - \frac{\alpha}{\varepsilon} + 1\right) \langle a \rangle.$$

Since $r_\varepsilon(\beta), r_\varepsilon(\alpha) \in L^\infty(\mathbb{R})$ (uniformly in ε), we finally obtain that

$$\lim_{\varepsilon \to 0} \int_\alpha^\beta a\left(\frac{x}{\varepsilon}\right) dx = (\beta - \alpha)\langle a \rangle.$$

Using the density of the step functions in $L^1(\mathbb{R})$, we conclude the proof of the lemma. □

Lemma 7.3 (First oscillation lemma). *Assume Ω to be a hypercube*[4] *in $\mathbb{R}^d$. For any $f \in C(\bar{\Omega}; C_\#(Y))$, it holds for $|Y| = 1$ that*

$$\lim_{\varepsilon \to 0} \int_\Omega f\left(x, \frac{x}{\varepsilon}\right) dx = \int_\Omega \int_Y f(x,y) dx dy.$$

Proof. This proof is taken from the diploma thesis by Neuss-Radu (1992). It does not cover the statement in full generality, but it highlights the main ideas behind why things work.

We set $i := (i_1, \ldots, i_d) \in \mathbb{Z}^d$, $Y_i = i + Y$, and $J := \{i \in \mathbb{Z}^d : \varepsilon Y_i \subset \Omega\}$.

For each $\delta > 0$, we can find an $\epsilon_0 > 0$ such that:

(1)

$$\int_{\Omega \setminus \cup_{i \in J} \varepsilon Y_i} f\left(x, \frac{x}{\varepsilon}\right) dx < \delta;$$

(2) $\|f(x,\cdot) - f(x',\cdot)\|_{C_\#(\bar{Y})} \leq \delta$ for all $|x - x'| < \varepsilon$ and for all $\varepsilon < \varepsilon_0$.

---

[4]This assumption on the geometry of Ω can be removed. We keep it here just for the sake of simplicity of the proofs.

We have that

$$\begin{aligned}\int_\Omega f\left(x, \frac{x}{\varepsilon}\right) dx &= \sum_{\varepsilon Y_i \subset \Omega} \int_{\varepsilon Y_i} f\left(x, \frac{x}{\varepsilon}\right) dx + \mathcal{O}(\delta)\\ &\overset{(2)}{=} \sum_{\varepsilon Y_i \subset \Omega} \int_{\varepsilon Y_i} f\left(x_i, \frac{x}{\varepsilon}\right) dx + \mathcal{O}(\delta)\\ &= \sum_{\varepsilon Y_i \subset \Omega} \varepsilon^d \int_{Y_i} f(x_i, y)\, dy + \mathcal{O}(\delta)\\ &= \int_\Omega \int_Y f(x,y) dx dy + \mathcal{O}(\delta).\end{aligned}$$

□

Lemma 7.4 (Second oscillation lemma). *Let the hypothesis of Lemma 7.3 hold. For all functions $g \in C(\bar{\Omega}; C_\#(\Gamma))$, we have that*

$$\lim_{\varepsilon \to 0} \varepsilon \int_{\Gamma_\varepsilon} g\left(x, \frac{x}{\varepsilon}\right) d\sigma_\varepsilon = \int_\Omega \int_\Gamma g(x,y) d\sigma_y dx.$$

A proof of this result can be found, for instance, in the work of Neuss-Radu (1992) and Allaire *et al.* (1995).

A quite useful tool in elucidating asymptotic limits is the following result.

Lemma 7.5. *Let $p \in [1, +\infty)$, $F_\varepsilon \in L^\infty(\Omega)$, with $F_\varepsilon \overset{*}{\rightharpoonup} \langle F \rangle$ in $L^\infty(\Omega)$ as $\varepsilon \to 0$, and $G_\varepsilon \in L^p(\Omega)$ such that $G_\varepsilon \to G_0$ strongly in $L^p(\Omega)$ as $\varepsilon \to 0$. Then,*

$$F_\varepsilon G_\varepsilon \rightharpoonup \langle F \rangle G_0 \text{ weakly in } L^1(\Omega) \quad \text{as } \varepsilon \to 0.$$

The reader may want to give a proof for this result.

Following the lines of Chapter 3 in Müller (1999), we note that a good feature for a macroscopic observable is that the upscaled behavior of a sequence, $u_\varepsilon : \Omega \to \mathbb{R}^d$, should be determined by the limit behavior of $\int_{\Omega'} f(u_\varepsilon)$ for continuous functions f and for any measurable sets Ω' from Ω. Denote by $C_0(\mathbb{R}^d)$ the closure of continuous functions on $\mathbb{R}^d$ with compact support. Then,

$$\mathcal{M}(\mathbb{R}^d) := (C_0(\mathbb{R}^d))^*$$

is seen here as the space of signed Radon measures with finite mass. This identification is done via the dual pairing

$$\langle \nu, f \rangle = \int_{\mathbb{R}^d} f d\nu.$$

Definition 7.2 (Weak-$\star$ measurability). The mapping $\nu : \Omega \to \mathcal{M}(\mathbb{R}^d)$ is weak-$\star$ measurable if, for all $f \in C_0(\mathbb{R}^d)$, the functions $x \to \langle \nu_x, f \rangle$ are measurable.

Here, ν_x denotes the measure $\nu(x)$. The following result, called the fundamental theorem for Young measures, gives a measure-theoretic description for the incompatibility between weak convergence and the composition with a continuous nonlinear function.

Theorem 7.3 (Young measures). *Let (u_ε) be a bounded sequence in $L^\infty(\Omega; \mathbb{R}^d)$. Then, there exist a subsequence (u_{ε_k}) of (u_ε) and, for a.e. $x \in \Omega$, a Borel probability measure ν_x on $\mathbb{R}^d$ such that for each $f \in C_0(\mathbb{R}^d)$, it holds*

$$f(u_{\varepsilon_k}) \overset{*}{\rightharpoonup} \bar{f} \quad \text{in } L^\infty(\Omega),$$

where

$$\bar{f}(x) := \int_{\mathbb{R}^d} F(z) d\nu_x(z) \quad \text{a.e. } x \in \Omega.$$

Here, $(\nu_x)_{x \in \Omega}$ is the family of Young measures generated by the sequence u_{ε_k}.

For further discussions regarding this result, we refer the reader to Theorem 11 in Evans (1990) and Theorem 3.1 in Müller (1999).

7.5 Compactness

Theorem 7.4 (Eberlein–Smuljan). *Let B be a reflexive Banach space, and let (u_ε) be bounded in B. Then, there exists a subsequence $(u_{\varepsilon_k}) \subset (u_\varepsilon)$ and a limit point $u \in B$ such that $u_{\varepsilon_k} \rightharpoonup u$ (weakly) in B as $k \to \infty$.*

See also Cioränescu and Donato (1999, Theorem 1.18).

As a rule, we denote the subsequence (u_{ε_k}) mentioned in Theorem 7.4 by (u_ε). Moreover, nearly everywhere in the textbook, whenever we take the limit as $\varepsilon \to 0$ with the sequence (u_ε), we actually mean (in the spirit of Theorem 7.4) that the limit function is reached only through a subsequence.

Theorem 7.5 (Rellich–Kondrachov). *The space $W^{1,p}(\Omega)$ is compactly embedded in $L^q(\Omega)$ for any $q \in [1, p^*)$, where $p^* = \frac{dp}{d-p}$ is the so-called critical Sobolev exponent. If $p^* > p$ and $d = p$ (hence $p^* = \infty$), then we have that $W^{1,p}(\Omega)$ is compactly embedded in $L^p(\Omega)$ for all $p \in [1, \infty]$.*

The case of $d < p \leq \infty$ is covered by the so-called Morrey's inequality. We refer to Evans (2015) for a proof of Theorem 7.5.

For instance, we often use the fact that $H^1(\Omega)$ is compactly embedded in $L^2(\Omega)$. This is a direct consequence of Theorem 7.5.

Lemma 7.6 (Schauder). *Let B_1 and B_2 be two Banach spaces. A bounded operator $T : B_1 \to B_2$ is compact if and only if the adjoint operator T^*: $B_2^* \to B_1^*$ is compact.*

See Brezis (2010, Theorem 6.4) for a proof.

Lemma 7.7 (Aubin–Lions). *Let $B_0 \subset B_1 \subset B_2$ be three Banach spaces. Denote by $S := (0, T)$, for some $T > 0$, a time interval and take $t \in S$. We assume that the embedding of B_1 into B_2 is continuous, while the embedding of B_0 into B_1 is compact. Let p, r such that $1 \leq p < +\infty$ and $1 \leq r \leq +\infty$. Then, the embedding of*

$$E_{p,r} := \left\{ v \in L^p(S; B_0) \;\middle|\; \frac{dv}{dt} \in L^r(S; B_2) \right\}$$

in $L^p(S; B_1)$ is compact, where the norm of $E_{p,r}$ is defined by

$$\|v\|_{E_{p,r}} := \|v\|_{L^p(S;B_0)} + \left\| \frac{dv}{dt} \right\|_{L^r(S;B_2)}.$$

See Boyer and Fabrie (2013, Theorem II.5.16) or Showalter (1997, Proposition III.1.3) for a proof.

7.6 Two-Scale Convergence: Definition, Properties, and Admissible Test Functions

In this section, we introduce the concept of two-scale convergence. This is a type of weak convergence which is adapted to sequences of periodic functions. The key observation behind the development of the concept of two-scale convergence is that we expect u_ε, the solution of a linear elliptic partial differential equation with oscillating coefficients as discussed in Section 2.3, to behave like

$$u_0(x) + \varepsilon u_1\left(x, \frac{x}{\varepsilon}\right).$$

Such asymptotic behavior is precisely what the formal asymptotic homogenization *ansatz* encodes in the proposed expansion. As formal asymptotic arguments turned out to be very efficient in guessing the structure of the upscaled equations and delivered in many cases explicit formulas for the effective coefficients, one expects that a type of convergence mimicking the homogenization *Ansatz* has a very good chance of being successful. Such ideas did lead G. Nguetseng to discover, in the late 1980s, a rigorous concept of a particular type of weak convergence that fits excellently to the type of sequences of periodic functions arising in homogenization problems; see Definition 7.3. For more details on this topic and on how G. Nguetseng and G. Allaire began to used it, we refer the reader to their original works (Nguetseng, 1989; Allaire, 1992) and other articles citing them. The basic methodology is also discussed in the excellently written article by Lukkassen *et al.* (2002). The two-scale concept has been developed further by A. Visintin, who even gave two-scale versions of standard theorems such as those of Sobolev, Rellich, and Morrey; see Visintin (2006). It is also worth noting that concepts similar to two-scale convergence have been adapted to work with measures or to be applicable in some stochastic cases as well; see for instance Lukkassen and Wall (2005) and Zhikov and Piatnitskii (2006). As a side note, the concept of Young measure can be adapted to cope with products of two-scale convergent sequences giving rise to multiscale Young measures, as given by Pedregal (2005).

Definition 7.3 (Two-scale convergence). Let (u_ε) be a sequence of functions in $L^2(\Omega)$. We say that (u_ε) converges two-scale to a unique function $u_0(x,y) \in L^2(\Omega \times Y)$ if and only if for all $\varphi \in C_0^\infty(\Omega, C_\#^\infty(Y))$, we have

$$\lim_{\varepsilon \to 0} \int_\Omega u_\varepsilon(x)\varphi\left(x, \frac{x}{\varepsilon}\right)dx = \frac{1}{|Y|}\int_\Omega \int_Y u_0(x,y)\varphi(x,y)dxdy. \tag{7.9}$$

We denote (7.9) by $u_\varepsilon \overset{2}{\rightharpoonup} u_0$. Often, we will assume $|Y| = 1$. This will be the case when we handle the homogenization of partial differential equations with oscillatory coefficients, which are posed in homogeneous domains.

Note that, in general, care needs to be paid whether setting $|Y| = 1$ makes sense physically and geometrically. It is often not suitable for the case when the domains are heterogeneous.[5]

We have seen that the first oscillation lemma (see Lemma 7.3) allows

$$\lim_{\varepsilon \to 0} \int_\Omega \varphi^2 \left(x, \frac{x}{\varepsilon}\right) dx = \int_{\Omega \times Y} \varphi^2(x, y) dx dy \tag{7.10}$$

for $\varphi \in C(\bar{\Omega}; C_\#(Y))$ and $|Y| = 1$.

Definition 7.4 (Admissible test function in Definition 7.3). A test function φ satisfying (7.10) is called *admissible.*

We point out briefly that there is a subtle issue concerning the choice of test functions for the two-scale convergence; see also the line of arguments made by Alouges (2016) and Lukkassen *et al.* (2002). We are often tempted to test with $\psi \in L^2(\Omega \times Y)$, but it can happen that the mapping

$$x \mapsto \varphi\left(x, \frac{x}{\varepsilon}\right)$$

is not Lebesgue measurable. A sufficient condition repairing this issue would be to assume that φ is a Carathéodory function, i.e. φ is continuous in one variable and measurable with respect to the second variable. This means that functions from sets like $L^2(\Omega, C_\#(Y))$ or $L^2_\#(Y, C(\bar{\Omega}))$ are such Carathéodory functions, and hence, if they also satisfy 7.10, then they are admissible as test functions in the two-scale convergence.

One of the main properties of the two-scale convergence is the convergence of the norms. We refer to this feature as the "continuity theorem", and state what we mean as follows.

Theorem 7.6 (Continuity theorem). *Let (u_ε) be a sequence in $L^2(\Omega)$, and let $u_0 \in L^2(\Omega \times Y)$ such that $u_\varepsilon \overset{2}{\rightharpoonup} u_0 \in L^2(\Omega \times Y)$. Then, we have the following properties:*

$$u_\varepsilon \rightharpoonup \langle u_0 \rangle_Y \text{ weakly in } L^2(\Omega)$$

[5] The use of $|Y| = 1$ can be very misleading especially in the case when the periodic cell is denoted differently, say by Z, and periodically placed microstructures are present (see the following chapter for more in this direction). $Y \subset Z$ then designates the microstructure, and usually, we have $|Y| \neq 1$.

and

$$\liminf_{\varepsilon\to 0} \|u_\varepsilon\|_{L^2(\Omega)} \geq \|u_0\|_{L^2(\Omega\times Y)}.$$

The notation $\langle\cdot\rangle_Y$ indicates that the average is taken on the set Y and not on the set $\Omega \times Y$. For a proof of this result, see e.g. Neuss-Radu (1992) and Alouges (2016).

The main result derived by Nguetseng (1989) and Allaire (1992) is the two-scale compactness result reported in Theorem 7.7. Such a compactness property makes the concept of convergence presented in Definition 7.3 a powerful tool.

Theorem 7.7 (Two-scale compactness). *The following statements hold:*

(i) *From each bounded sequence* (u_ε) *in* $L^2(\Omega)$, *one can extract a subsequence which two-scale converges to* $u_0(x,y) \in L^2(\Omega \times Y)$.
(ii) *Let* (u_ε) *be a bounded sequence in* $H^1(\Omega)$, *which converges weakly to a limit function,* $u_0 \in H^1(\Omega)$. *Then, there exists* $u_1 \in L^2(\Omega; H^1_\#(Y)/\mathbb{R})$, *such that up to a subsequence,* (u_ε) *two-scale converges to* $u_0(x,y)$ *as before and* $\nabla u_\varepsilon \overset{2}{\rightharpoonup} \nabla_x u_0 + \nabla_y u_1$.
(iii) *Let* (u_ε) *and* $(\varepsilon\nabla u^\varepsilon)$ *be bounded sequences in* $L^2(\Omega)$. *Then, there exists* $u_0 \in L^2(\Omega; H^1_\#(Y))$ *such that, up to a subsequence,* u_ε *and* $\varepsilon\nabla u_\varepsilon$ *two-scale converge to* $u_0(x,y)$ *and* $\nabla_y u_0(x,y)$, *respectively.*

We are particularly interested in treating the presence of a periodic array of inclusions (perforations, microstructures, ...). Consequently, a concept of two-scale convergence similar to the one from Definition 7.3 is needed to deal with periodically placed material surfaces – the boundary of a periodic array of inclusions.

Definition 7.5 (Two-scale convergence for ϵ-periodic surfaces; cf. Allaire *et al.* (1995) and Neuss-Radu (1996)). A sequence of functions (u_ε) in $L^2(\Gamma_\varepsilon)$ is said to two-scale converge to a limit $u_0 \in L^2(\Omega \times \Gamma)$ if and only if for any $\varphi \in C_0^\infty(\Omega, C_\#^\infty(\Gamma))$, we have

$$\lim_{\varepsilon\to 0} \epsilon \int_{\Gamma_\varepsilon} u^\varepsilon(x)\varphi\Big(x, \frac{x}{\epsilon}\Big) d\sigma_\varepsilon = \int_\Omega \int_\Gamma u_0(x,y)\varphi(x,y) d\sigma_y dx.$$

The concern about what test function is admissible and what is not for the concept of two-scale convergence adapted for oscillating surfaces is present here as well. We have seen that the second oscillation lemma (see Lemma 7.4) allows

$$\lim_{\varepsilon\to 0}\int_{\Gamma_\varepsilon}\varphi^2\left(x,\frac{x}{\varepsilon}\right)d\sigma_\varepsilon = \frac{1}{|Y|}\int_{\Omega\times\Gamma}\varphi^2(x,y)dxd\sigma_y \tag{7.11}$$

for $\varphi \in C(\bar{\Omega}; C_\#(\Gamma))$ and $|Y| = 1$.

Definition 7.6 (Admissible test function in Definition 7.5). A test function φ satisfying (7.11) is called *admissible*.

Functions from sets such as $L^2(\Omega, C_\#(\Gamma))$ or $L^2_\#(\Gamma, C(\bar{\Omega}))$ are Carathéodory functions, and hence, they are admissible as test functions in the two-scale convergence for oscillating surfaces; cf. Definition 7.5.

Theorem 7.8 (Two-scale compactness for oscillating surfaces). *The following statements hold:*

(i) *From each bounded sequence* $(u_\varepsilon) \in L^2(\Gamma_\varepsilon)$*, one can extract a subsequence,* u^ε*, which two-scale converges to a function,* $u_0 \in L^2(\Omega \times \Gamma)$*.*
(ii) *If a sequence of functions,* (u_ε)*, is bounded in* $L^\infty(\Gamma_\varepsilon)$*, then* u_ε *two-scale converges to a function,* $u_0 \in L^\infty(\Omega \times \Gamma)$*.*

Proof. For the proof of (i), see Neuss-Radu (1996). The proof of (ii) is given, for instance, by Marciniak-Czochra and Ptashnyk (2008). □

The two-scale convergence concepts introduced in this section are applicable to evolution problems, provided a few elementary modifications are made to cope with the time variable; see e.g. Allaire (1992), Lukkassen *et al.* (2002), and Lukkassen and Wall (2005). Essentially, we extend the definitions to time-dependent functions by regarding time as a parameter; hence, we do not take into account oscillations in the variable t and work with the time-space cylinder $\Omega_T := (0,T) \times \Omega$. Without giving any specific details, we present here the main concepts by exploiting the analogy with the time-independent case.

Definition 7.7. A sequence (u_ε) in $L^2(\Omega_T)$ two-scale converges to a function $u_0 \in L^2(\Omega_T; L^2(Y))$ if there exists at least a subsequence, denoted again by (u_ε), such that

$$\lim_{\varepsilon\to 0}\int_{\Omega_T} u_\varepsilon(t,x)\varphi(t,x,\frac{x}{\varepsilon})dtdx = \frac{1}{|Y|}\int_{\Omega_T}\int_Y u_0(t,x,y)\varphi(t,x,y)dtdxdy$$

for all test functions $\varphi \in C_0^\infty(\Omega_T; C_\#^\infty(Y))$.

One should imagine the test functions $\varphi \in C_0^\infty(\Omega_T; C_\#^\infty(Y))$ as having the structure $\varphi(t,x,y) = \psi(x,y)\eta(t)$, with $\psi \in C_0^\infty(\Omega; C_\#^\infty(Y))$ (admissible, as discussed previously) and $\eta \in C_0^\infty(0,T)$.

In the same spirit, the two-scale compactness results extend here as well. Hence, from any uniformly bounded sequence (u_ε) in $L^2(\Omega_T)$, we can extract a subsequence such that

$$u_\varepsilon \overset{2}{\rightharpoonup} u_0 \in L^2(\Omega_T \times Y).$$

Furthermore, if (u_ε) is uniformly bounded in $H^1(\Omega_T)$, then there exist $u_0 \in L^2(\Omega_T \times Y)$ (as before) and $u_1 \in L^2(\Omega_T; H^1(Y)/\mathbb{R})$ such that

$$\nabla u_\varepsilon \overset{2}{\rightharpoonup} \nabla_x u_0 + \nabla_y u_1.$$

Definition 7.8. A sequence (u_ε) in $L^2(0,T) \times \Gamma_\varepsilon)$ two-scale converges to a function $u_0 \in L^2(\Omega_T; L^2(\Gamma))$ if

$$\lim_{\varepsilon\to 0} \varepsilon \int_0^T \int_{\Gamma_\varepsilon} u_\varepsilon(t,x)\varphi\left(t,x,\frac{x}{\varepsilon}\right) d\sigma_\varepsilon dt = \frac{1}{|Y|}\int_{\Omega_T}\int_\Gamma u_0(t,x,y)\varphi(t,x,y)dtdxd\sigma_y$$

holds for all test functions $\varphi \in C_0^\infty(\Omega_T; C_\#^\infty(\Gamma)$.

The test functions $\varphi \in C_0^\infty(\Omega_T; C_\#^\infty(\Gamma))$ have the structure $\varphi(t,x,y) = \psi(x,y)\eta(t)$, with $\psi \in C_0^\infty(\Omega; C_\#^\infty(\Gamma))$ (admissible) and $\eta \in C_0^\infty(0,T)$.

Regarding the corresponding compactness result for time-dependent oscillating sequences defined on hypersurfaces, we simply note that from any uniformly bounded sequence (u_ε) in $L^2(0,T; L^2(\Gamma_\varepsilon))$, we can extract a subsequence such that, as $\varepsilon \to 0$, the following convergence holds:

$$u_\varepsilon \overset{2}{\rightharpoonup} u_0 \in L^2(\Omega_T \times \Gamma).$$

Depending on the concrete application at hand, various types of multiscale convergence might be needed. For instance, if slight deviations from periodicity intervene, then instead of (7.9) one could have a convergence tailored to handle locally periodic scenarios (Ptashnyk, 2013) or weakly stochastic ones (Heida, 2011). If convection dominates diffusion for problems posed in periodic media, then the right convergence tool seems to be a variant of (7.9), called two-scale convergence with drift (Allaire *et al.*, 2010), concepts that can be related to the study of the classical Taylor dispersion. Not all multiscale effects can be captured using the concept of two-scale convergence. This is the case, for example, when localized defects appear in periodic settings; see, e.g. Wolf (2022).

7.7 Exercises

Exercise 7.7.1. Show that $H^1_\#(Y)$ and $L^2(\Omega; H^1_\#(Y))$ are Hilbert spaces.

Exercise 7.7.2. Is the embedding $L^2(\Omega; H^1_\#(Y))$ into $L^2(\Omega; L^2_\#(Y))$ compact? Justify your answer.

Exercise 7.7.3. Prove Theorem 7.1 using Banach's fixed-point principle.

Exercise 7.7.4. Let $\Omega \subset \mathbb{R}^d$ be a bounded and smooth domain. We consider the boundary-value problem

$$\begin{aligned} \operatorname{div}(-A\nabla u) + \alpha u &= 0 \quad \text{in } \Omega, \\ -A\nabla u \cdot n &= \beta(u-1) \quad \text{on } \partial\Omega, \end{aligned}$$

where $A \in C(\overline{\Omega})$ is scalar and positive, $\alpha, \beta > 0$, and n is the normal vector of $\partial\Omega$ pointing outward.

(i) Write down a weak form for this boundary-value problem.
(ii) Show the existence of a unique solution via the Lax–Milgram lemma.

Exercise 7.7.5. Let $\Omega \subset \mathbb{R}^d$ be a bounded, smooth domain. Set $\delta > 0$, and consider the functions $\alpha, \beta \in L^\infty(\Omega)$, and $f \in L^2(\Omega)$. Given the boundary-value problem

$$\begin{aligned} -\delta\Delta u + \operatorname{div}(\beta u) + \alpha u &= f \quad \text{in } \Omega, \\ u &= 0 \quad \text{on } \partial\Omega, \end{aligned}$$

perform the following tasks:

(i) Write down a weak form for this boundary-value problem.
(ii) Additionally, assume

$$\delta\|v\|^2_{H^1_0(\Omega)} + 2\int_\Omega \left(-\beta(x)\cdot\nabla v(x)v(x) + \alpha(x)v(x)^2\right) dx \geq 0$$

for all $v \in H^1_0(\Omega)$. Use the Lax–Milgram lemma to show that there exists a unique solution.
(iii) Set $\Omega = (0,1)^d$. Try to find a non-trivial[6] choice for the functions α, β such that the above condition is satisfied for all $v \in H^1_0(\Omega)$.

[6] Both functions are not zero. However, the use of the constant function is allowed.

(iv) What do you expect to happen with u for $\delta \to 0$? Use your intuition – no rigorous proof is required at this stage.

Exercise 7.7.6. Prove Lemma 7.5.

Exercise 7.7.7. Show that if $u_\varepsilon \to u$ strongly in $L^2(\Omega)$, then the following statements hold:

(i) $u_\varepsilon \rightharpoonup u$ in $L^2(\Omega)$;
(ii) $u_\varepsilon \overset{2}{\rightharpoonup} u_0(x,y)$ in $L^2(\Omega \times Y)$, where $u_0(x,y) = u(x)$ for $(x,y) \in \Omega \times Y$.

Exercise 7.7.8. Does the two-scale convergence in $L^2(\Omega \times Y)$ imply weak convergence in $L^2(\Omega)$? Show your arguments.

Exercise 7.7.9. Let $u : \mathbb{R} \to [-1,1]$ be defined by $u(y) = \cos 2\pi y$. Set $u_\varepsilon(x) = u\left(\frac{x}{\varepsilon}\right)$ for $x \in (a,b)$, where $a, b \in \mathbb{R}$ and $a < b$.

(i) Show that $u_\varepsilon \rightharpoonup 0$ in $L^2(a,b)$ as $\epsilon \to 0$.
(ii) Can one also have $u_\varepsilon \to 0$ in $L^2(a,b)$? Justify your answer.
(iii) We expect that $u_\varepsilon \overset{2}{\rightharpoonup} u_0$ in $L^2(\Omega \times Y)$. Can you guess which are the best candidates for u_0 and Y? Justify your answer.

Exercise 7.7.10. Consider the domain $\Omega = (-1,1)$ and the function $u_\varepsilon : \Omega \to \mathbb{R}$, where

$$u_\varepsilon(x) = \begin{cases} \varepsilon, & \text{if } x \in \left(0, \dfrac{1}{\varepsilon}\right), \\ 0, & \text{if } x \in \mathbb{R} \backslash \left(0, \dfrac{1}{\varepsilon}\right). \end{cases}$$

Show that the sequence (u_ε) is bounded in $L^1(\Omega)$, and calculate its weak limit.

Exercise 7.7.11. Give conditions so that the product of two admissible test functions (as needed in Definition 7.3, or, respectively, in Definition 7.5) is also admissible. Justify your answer.

Exercise 7.7.12. Prove that the function $\varphi(x,y) = \zeta(x)\rho(y)$, where $\zeta \in L^2(\Omega)$ and $\rho \in L^2_\#(Y)$, is an admissible test function in the sense of Definition 7.10. Similarly, prove that the function $\psi(x,y) = \zeta(x)\eta(y)$, where $\zeta \in L^2(\Omega)$ and $\eta \in L^2_\#(\Gamma)$, is an admissible test function in the sense of Definition 7.6.

Exercise 7.7.13. Take $u_0(x,y)$ as admissible function in the sense of Definition 7.10. Prove that the function $u_\varepsilon(x) = u_0\left(x, \frac{x}{\varepsilon}\right)$ converges to u_0. Take now $v_\varepsilon(x) = u_0\left(x, \frac{x}{\varepsilon^3}\right)$. Does v_ε have a two-scale limit? Justify your answer.

Exercise 7.7.14. Formulate and prove a "continuity theorem" (similar to Theorem 7.6) that holds for the two-scale convergence for oscillating surfaces.

7.8 Solutions

Solution 7.7.3. In a typical homogenization context, the interest in Theorem 7.1 is motivated by the need for ensuring the existence and uniqueness of solutions to equations of type $Au = f$, where $u \in H$, where the operator $A : H \to H$ with $u \mapsto Au$ is a linear elliptic operator for all $u \in H$.

Essentially, we interpret $\mathcal{A}(v_1, v_2) := (Av_1, v_2)$ and $\mathcal{F}(v_2) = (f, v_2)$ for all $v_1, v_2 \in H$. For most applications we have in mind here, we can take $H := H_0^1(\Omega)$ and $f \in L^2(\Omega) \subset H$.

We define the fixed-point operator $T : H \to H$ by $T(u) := u - \rho(Au - f)$, where $\rho > 0$ is a parameter to be specified later. We only sketch here a proof showing that this operator is contractive for suitable values of ρ. This property can be obtained via a direct calculation as follows.

For any $u_1, u_2 \in H$, we have

$$\begin{aligned}
||T(u_1) - T(u_2)||_H^2 &= ||u_1 - \rho(Au_1 - f) - u_2 + \rho(Au_2 - f)||_H^2 \\
&= ||u_1 - u_2||_H^2 + \rho^2||A(u_1 - u_2)||_H^2 \\
&\quad - \rho\left[(u_1 - u_2, A(u_1 - u_2)) + (A(u_1 - u_2), u_1 - u_2)\right] \\
&= ||u_1 - u_2||_H^2 + \rho^2||A(u_1 - u_2)||_H^2 \\
&\quad - 2\rho\mathcal{A}(u_1 - u_2, u_1 - u_2) \\
&\leq ||u_1 - u_2||_H^2 + \rho^2||A(u_1 - u_2)||_H^2 - 2\alpha\rho||u_1 - u_2||_H^2 \\
&\leq ||u_1 - u_2||_H^2 \left(1 + \rho^2||A|| - 2\rho\alpha\right),
\end{aligned}$$

where $||A||$ denotes the norm of the operator A. In the last inequality, we used the coercivity of A. Now, we see that the boundedness of A is also needed to close our argument. If one takes $\rho > 0$ such that $\rho\left(\rho||A|| - 2\alpha\right) \in (-1, 0)$, then T becomes a contraction. Hence, there exists $u^\star \in H$ such that $T(u^\star) = u^\star$, i.e. $u^\star$ is the (unique) fixed point of T. Consequently, $u^\star$ must also satisfy $Au^\star = f$.

The proof of (7.2) follows by using the standard energy estimate.

Solution 7.7.4. It is advisable that you make use of the trace inequality: There exists $C > 0$ such that

$$\|u\|_{L^2(\partial\Omega)} \leq C\|u\|_{H^1(\Omega)}$$

for all $u \in H^1(\Omega)$.

Solution 7.7.7. Concerning point (ii), use the first oscillation lemma (Lemma 7.3), in combination with the continuity theorem (Theorem 7.6).

Solution 7.7.9. To give a rigorous answer to point (iii) of this exercise, one can use the first oscillation lemma (Lemma 7.3), part (i) of the two-scale compactness result (Theorem 7.7) in combination with the relation between two-scale convergence and weak convergence.

Solution 7.7.12. The first part of this exercise is taken from Exercise 2.1 (part 2) in Section 2.5 from Alouges (2016). The second part is a natural adaptation to the case of two-scale convergence for oscillating surfaces.

Appendix

FreeFEM++ Code Solving an Exercise from Chapter 6

About the Exercise from Chapter 6

Start with installing FreeFEM++ on your computer. As an alternative to the FEniCS example discussed in Chapter 6, you can use FreeFEM++ along with the codes listed here to build your own computational homogenization tool. Modify the following lines according to your preferred selection of parameters and geometry.

The file main.edp

```
// Declaration of macros
macro grad(u) [dx(u), dy(u)] //
macro av(u) [a1*dx(u)+a2*dy(u),a2*dx(u)+a3*dy(u)] //
macro aw(u) [a1e*dx(u)+a2e*dy(u),a2e*dx(u)+a3e*dy(u)] //
macro ac1 [a1e, a2e] //
macro ac2 [a2e, a3e] //

// Parameters for P_0
func g = 1;
func f0 = 2 + 2*sin(x*2*pi);

func a1e = 2 + 1*((sin(x*2*pi)*sin(y*2*pi)));
func a2e = 0 ;
func a3e = 2 + 1*((sin(x*2*pi)*cos(y*2*pi)));

```

```
include "hmgp0.edp" //execute P_0 solver (hmg)

for (int i=1;i<11;i++){
// Parameters for P_eps
int pore = i;
real eps = 1./pore;

func f = f0 + eps * (sin(2*pi*(x/eps)) * cos(2*pi*(y/eps)));
func a1 = 2 + 1*((sin((x*2*pi)/eps)*sin((y*2*pi)/eps)));
func a2 = 0;
func a3 = 2 + 1*((sin((x*2*pi)/eps)*cos((y*2*pi)/eps)));


include "pbeps.edp" //execute P_eps solver
plot(u, fill=true, value=true, cmm="A",wait=1);
plot(ue, fill=true, value=true, cmm="A",wait=1);
}
```

The file sqr.edp

```
// mesh quality parameters
int nSq = 100;
int[int] nnSq = [nSq, nSq, nSq, nSq];

// parameters for the outer square domain
real[int] xxSq = [0, 1, 1, 0],
          yySq = [0, 0, 1, 1];

// creating outer boundary
border bbSq(t=0, 1; i)
{
    int iiSq = (i+1)%4;
    real t1Sq = 1-t;
    x = xxSq[i]*t1Sq + xxSq[iiSq]*t;
    y = yySq[i]*t1Sq + yySq[iiSq]*t;
    label = 0;
}

// plot(bbSq(nnSq), wait=1);

// creating mesh
mesh Th = buildmesh(bbSq(nnSq));
// plot(Th, wait=1);
```

The file hmgp0.edp

```
//Domain
include "cell.edp"
include "sqr.edp"

fespace Wh(Thc,P1,periodic=[[0, x], [2, x], [1, y], [3, y]]);
Wh w1, w2, wv1, wv2;

// Cell Problem
problem cellP1 (w1, wv1)
   = int2d(Thc)(
        (aw(w1)' * grad(wv1))
   )
     + int2d(Thc)(
        (ac1' * grad(wv1))
   );
problem cellP2 (w2, wv2)
   = int2d(Thc)(
        (aw(w2)' * grad(wv2))
   )
     + int2d(Thc)(
        (ac2' * grad(wv2))
   );

cellP1;
cellP2;

//A_eff
real D1 = (int2d(Thc)((a1e+(ac1'*grad(w1)))));
real D2 = (int2d(Thc)((a2e+(ac1'*grad(w2)))));
real D3 = (int2d(Thc)((a2e+(ac2'*grad(w1)))));
real D4 = (int2d(Thc)((a3e+(ac2'*grad(w2)))));


real GNI = int1d(Thc,4)(g);

macro ay(u) [D1*dx(u)+D2*dy(u),D3*dx(u)+D4*dy(u)] //

fespace Vh(Th,P1);
Vh ue, ve;

```

```
// Homogenized problem P0
problem homP (ue, ve)
   = int2d(Th)(
      (ay(ue)' * grad(ve))
   )
    + int2d(Th)(
      GN * (ue * ve)
   )
   - int2d(Th)(
      (f0 * ve)
   )
   + on(0, ue=0)
   ;
homP;

plot(w1, fill=true, value=true, cmm="A",wait=1);
```

The file cell.edp

```
// number of perforations along 1 axis
int cnlCell = 1;

// number of perforations in the whole domain
int cnCell = cnlCell*cnlCell;

// mesh quality parameters
int nCell = 100;
int niCell = -nCell / cnlCell;

// parameters for the outer square domain
real[int] xxCell = [0, 1, 1, 0],
          yyCell = [0, 0, 1, 1];

// ridius and center's coordinates in the reference cell
real riCell = 0.166;
real xiCell = 0.5;
real yiCell = 0.5;
real GN = 2*riCell*pi;

// generating parameters for every perforation in the domain
int myCell = 0;
int mxCell;
real[int] RCCell(cnCell);
```

```
int[int] NCCell(cnCell);
real[int] XCCell(cnCell);
real[int] YCCell(cnCell);
RCCell[0] = riCell;
NCCell[0] = niCell;
XCCell[0] = xiCell;
YCCell[0] = yiCell;

// creating outer boundary
real l0 = 0, r0 = 1, b0 = 0, t0 = 1;
border Gb0(t = l0, r0) {x = t; y = b0; label = 0;}
border Gl0(t = b0, t0) {x = r0; y = t; label = 1;}
border Gt0(t = r0, l0) {x = t; y = t0; label = 2;}
border Gr0(t = t0, b0) {x = l0; y = t; label = 3;}

// creating perforations
border ccCell(t=0, 2*pi; i)
{
    x = RCCell[i]*cos(t) + XCCell[i];
    y = RCCell[i]*sin(t) + YCCell[i];
    label = 4;
}

// plot(Gb0(nCell), Gl0(nCell), Gt0(nCell), Gr0(nCell), ccCell(niCell), wait=1);
// creating mesh
mesh Thc = buildmesh(Gb0(nCell)+Gl0(nCell)+Gt0(nCell)+Gr0(nCell)+ccCell(niCell),fixedborder=true);
// plot(Thc, wait=1);
```

The file *dompeps.edp*

```
// number of perforations in the whole domain
int cn = pore*pore;

// mesh quality parameters
int n = 100;
int[int] nn = [n, n, n, n];
int ni = (-n / pore) -5 ;

// parameters for the outer square domain
real[int] xx = [0, 1, 1, 0],
          yy = [0, 0, 1, 1];

```

```
// ridius and center's coordinates in the reference cell
real ri = 0.25;
real xi = 0.5;
real yi = 0.5;

// ridius and center's coordinates in the domain
real rf = ri/pore;
real xf = xi/pore;
real yf = yi/pore;

// generating parameters for every perforation in the domain
int my = 0;
int mx;
real[int] RC(cn);
int[int] NC(cn);
real[int] XC(cn);
real[int] YC(cn);
for (int i = 0; i < cn; i++){
    RC[i] = rf;
    NC[i] = ni;
    if (i % pore == 0){
        my = my + 2;
        mx = 2;};
    XC[i] = mx*xf - xf;
    YC[i] = my*yf - yf;
    mx = mx + 2;
}

// creating outer boundary
border bb(t=0, 1; i)
{
    int ii = (i+1)%4;
    real t1 = 1-t;
    x = xx[i]*t1 + xx[ii]*t;
    y = yy[i]*t1 + yy[ii]*t;
    label = 0;
}

// creating perforations
border cc(t=0, 2*pi; i)
{
    x = RC[i]*cos(t) + XC[i];
    y = RC[i]*sin(t) + YC[i];
    label = 1;
}
```

```

// plot(bb(nn), cc(NC), wait=1);

// creating mesh
mesh Thp = buildmesh(bb(nn) + cc(NC));
// plot(Thp, wait=1);
```

The file pbeps.edp

```
//Domain
include "dompeps.edp"

fespace Uh(Thp,P1);
Uh u, v;

// Direct problem P_eps
problem vP (u, v)
   = int2d(Thp)(
       (av(u)' * grad(v))
   )
     + int2d(Thp)(
        sqrt(eps) *( u * v)
   )
     + int1d(Thp,1)(
        eps *( u * v)
   )
   - int2d(Thp)(
        (f * v)
   )
   + on(0, u=0)
   ;

vP;
// Plot
// plot(u, fill=true, value=true, cmm="A",wait=1);
```

Bibliography

Abdulle, A. and Nonnenmacher, A. (2009). "A short and versatile finite element multiscale code for homogenization problems," *Computer Methods in Applied Mechanics and Engineering* **198**, 37, pp. 2839–2859.

Acerbi, E., Chiado Piat, V., Dal Maso, G., and Percivale, D. (1992). "Two-scale convergence with respect to measures and homogenization of monotone operators," *Nonlinear Analysis, Theory, Methods & Applications* **18**, pp. 481–496.

Adams, R. A. and Fournier, J. F. (2003). *Sobolev Spaces* (Academic Press, New York).

Aiki, T., Kröger, N. H., and Muntean, A. (2021). "A macro-micro elasticity-diffusion system modeling absorption-induced swelling in rubber foams: Proof of the strong solvability," *Quarterly of Applied Mathematics* **79**, pp. 545–579.

Ainouz, A. (2007). "Two-scale homogenization of a Robin problem in perforated media," *Applied Mathematical Sciences* **36**, pp. 1789–1802.

Aleksanyan, H., Shahgholian, H., and Sjölin, P. (2013). "Applications of Fourier analysis in homogenization of Dirichlet problem I. Pointwise estimates," *Journal of Differential Equations* **256**, pp. 2626–2637

Alexandrian, A. (2015). "A primer on homogenization of elliptic pdes with stationary and ergodic random coefficient functions," *Rocky Mountain Journal of Mathematics* **45**, 2, pp. 703–735.

Allaire, G. (1992). "Homogenization and two-scale convergence," *SIAM Journal on Mathematical Analysis* **23**, 6, pp. 1482–1518.

Allaire, G. (2002a). "Introduction to homogenization. Lecture notes of a summer school," ICTP, Trieste.

Allaire, G. (2002b). *Shape Optimization by the Homogenization Method* (Springer Verlag, New York).

Allaire, G. (2012). "A brief introduction to homogenization and miscellaneous applications," *ESAIM: Proceedings* **47**, pp. 1–49.

Allaire, G., Damlamian, A., and Hornung, G. (1995). "Two-scale convergence on periodic surfaces and applications," in A. Bourgeat, C. Carasso, S. Luckhaus, and A. Mikelic (eds.), *Proceedings of the International Conference on Mathematical Modeling of Flow through Porous Media* (World Scientific, Singapore), pp. 15–25.

Allaire, G., Mikelic, A., and Piatnitskii, A. (2010). "Homogenization approach to the dispersion theory for reactive transport through porous media," *SIAM Journal on Mathematical Analysis* **42**, pp. 125–144.

Allaire, G. and Murat, F. (1993). "Homogenization of the Neumann problem with nonisolated holes," *Asymptotic Analysis* **7**, 2, pp. 81–95.

Alnaes, M., Blechta, J., Hake, J., Johansson, A., Kehlet, B., Logg, A., Richardson, C., J., R., Rognes, J., and Wells, G. (2015). "The FEniCS project version 1.5," *Archive of Numerical Software* **3**, 100, pp. 9–23.

Alouges, F. (2016). "Introduction to periodic homogenization," *Interdisciplinary Information Sciences* **22**, 2, pp. 147–186.

Amar, M., Andreucci, D., and Cirillo, E. N. (2025). "Upscaled equations for the Fokker–Planck diffusion through arrays of permeable and of impermeable inclusions," *Asymptotic Analysis* **142**, 3, doi:10.1177/09217134241308419.

Aubin, J.-P. (1963). "Un théorème de compacité," *Comptes Rendus de l'Académie des Sciences, Paris* **256**, pp. 5042–5044.

Auriault, J. L. (1991). "Heterogeneous medium. is an equivalent macroscopic description possible?" *International Journal of Engineering Science* **29**, pp. 785–795.

Bakhalov, N. and Panasenko, G. (1989). *Homogenisation: Averaging Processes in Periodic Media. Mathematical Problems in the Mechanics of Composite Materials*, Mathematics and its Applications, Vol. 36 (Kluwer Academic Publishers, Dordrecht, Netherlands).

Barenblatt, G. I. (2003). *Scaling*, Cambridge Texts in Applied Mathematics, Vol. 34 (Cambridge University Press, Cambridge).

Bear, J. (1988). *Dynamics of Fluids in Porous Media* (Dover, New York).

Bensoussan, A., Lions, J.-L., and Papanicolaou, G. (1978). *Asymptotic Analysis for Periodic Structures* (American Mathematical Society, Providence, RI).

Berlyand, L. and Rybalko, V. (2018). *Getting Acquainted with Homogenization and Multiscale*, Compact Textbooks in Mathematics (Birkhäuser, Basel, Switzerland).

Blanc, X., Le Bris, C., and Lions, P. L. (2012). "A possible homogenization approach for the numerical simulation of periodic microstructures with defects," *Milan Journal of Mathematics* **80**, pp. 351–362.

Bonetti, E., Cavaterra, C., Natalini, R., and Solci, M. (eds.) (2021). *Mathematical Modeling in Cultural Heritage (MACH 21), INdAM Series*, Vol. 41 (Springer Verlag, Cham, Switzerland).

Bowen, R. M. (1980). "Incompressible porous media models by use of the theory of mixtures," *International Journal of Engineering Science* **18**, pp. 1129–1148.

Boyer, F. and Fabrie, P. (2013). *Mathematical Tools for the Study of the Incompressible Navier-Stokes Equations and Related Models* (Springer Verlag, Cham, Switzerland).

Braga, A. G., Furtado, F., Moreira, J. M., and Rolla, L. (2003). "Renormalization group analysis of nonlinear diffusion equations with periodic coefficients," *Multiscale Modeling and Simulation* **4**, pp. 630–644.

Brezis, H. (2010). *Functional Analysis, Sobolev Spaces and Partial Differential Equations* (Springer Verlag, Cham, Switzerland).

Cannon, J. R. and Hill, C. D. (1970). "On the movement of a chemical reaction interface," *Indiana University Mathematics Journal* **20**, pp. 429–454.

Chapman, J., Hewett, D. P., and Trefethen, L. N. (2015). "Mathematics of the Faraday cage," *SIAM Review* **57**, 3, pp. 398–417.

Chechkin, G. A. (2021). "The Meyers estimates for domains perforated along the boundary," *Mathematics* **9**, p. 3015.

Chechkin, G. A., Piatnitski, A. L., and Shamaev, A. S. (2007). *Homogenization Methods and Applications* (American Mathematical Society, Providence, RI).

Chipot, M. (2002). *ℓ goes to plus infinity*, Basler Lehrbücher (Springer, Basel, Switzerland).

Ciorănescu, D. and Donato, P. (1999). *An Introduction to Homogenization*, Vol. 17 (Oxford University Press, Oxford).

Ciorănescu, D. and Saint Jean Paulin, J. (1979). "Homogenization in open sets with holes," *Journal of Mathematical Analysis and Applications* **71**, 2, pp. 590–607.

Ciorănescu, D. and Saint Jean Paulin, J. (1998). *Homogenization of Reticulated Structures*, Applied Mathematical Sciences, Vol. 136 (Springer Verlag, Cham, Switzerland).

Cirillo, E., Colangeli, M., Muntean, A., and Thieu, T. T. (2020). "When diffusion faces drift: Consequences of exclusion processes for bi-directional pedestrian flows," *Physica D: Nonlinear Phenomena* **413**, p. 132651.

Cirillo, E. and Muntean, A. (2013). "Dynamics of pedestrians in regions with no visibility — a lattice model without exclusion," *Physica A: Statistical Mechanics and Its Applications* **392**, 17, pp. 3578–3588.

Courard, L., Zhao, Z., and Michel, F. (2021). "Influence of hydrophobic product nature and concentration on carbonation resistance of cultural heritage concrete buildings," *Cement and Concrete Composites* **115**, p. 103860.

Crooks, E. C. M., Dancer, E. N., Hilhorst, D., Mimura, M., and Ninomiya, H. (2004). "Spatial segregation limit of a competition-diffusion system with Dirichlet boundary conditions," *Nonlinear Analysis: Real World Applications* **5**, pp. 645–665.

Dancer, E. N., Hilhorst, D., Mimura, M., and Peletier, L. A. (1999). "Spatial segregation limit of a competition-diffusion system," *European Journal of Applied Mathematics* **10**, pp. 97–115.

Davit, Y., C, G. B., Byrne, H. M., Chapman, L. A., Kimpton, L. S., Lang, G. E., Leonard, K. H., Oliver, J. M., Pearson, N. C., Shipley, R. J., Waters, S. L., Whiteley, J. P., Wood, B. D., and Quintard, M. (2013). "Homogenization via formal multiscale asymptotics and volume averaging: How do the two techniques compare?" *Advances in Water Resources* **62**, pp. 178–206.

DiBenedetto, E. (1993). *Degenerate Parabolic Equations* (Springer, New York).

Du, Q., Engquist, B., and Tian, X. (2020). "Multiscale modeling, homogenization and nonlocal effects," *Contemporary Mathematics* **754**, pp. 115–139.

Eden, M. and Muntean, A. (2017). "Corrector estimates for the homogenization of a two-scale thermoelasticity problem with a priori known phase transformations," *Electronic Journal of Differential Equations* **57**, pp. 1–21.

Evans, L. C. (1982). "A chemical diffusion-reaction free boundary problem," *Nonlinear Analysis: Theory, Methods & Applications* **6**, 5, pp. 455–466.

Evans, L. C. (1990). *Weak Convergence Methods for Nonlinear Partial Differential Equations*, CBMS - Regional Conference Series in Mathematics, Vol. 74 (American Mathematical Society, Providence, RI).

Evans, L. C. (2015). *Introduction to PDEs*, Textbooks in Mathematics, Vol. 23, 4th edn. (American Mathematical Society, Providence, RI).

Evers, J., Hille, S., and Muntean, A. (2015). "Mild solutions to a measure-valued mass evolution problem with flux boundary conditions," *Journal of Differential Equations* **259**, 3, pp. 1068–1097.

Fasano, A., Primicerio, M., and Ricci, R. (1990). "Limiting behaviour of some problems in diffusive penetration," *Rendiconti di Matematica* **10**, pp. 39–57.

Fatima, T., Arab, N., Zemskov, E., and Muntean, A. (2011). "Homogenization of a reaction-diffusion system modeling sulfate corrosion of concrete in locally periodic perforated domains," *Journal of Engineering Mathematics* **69**, 2, pp. 261–276.

Fatima, T., Ijioma, E., Ogawa, T., and Muntean, A. (2014). "Homogenization and dimension reduction of filtration combustion in heterogeneous thin layers," *Networks and Heterogeneous Media* **9**, 4, pp. 709–737.

Fatima, T. and Muntean, A. (2014). "Sulfate attack in sewer pipes: Derivation of a concrete corrosion model via two-scale convergence," *Nonlinear Analysis: Real World Applications* **15**, 1, pp. 326–344.

Friedman, A. and Tzavaras, A. T. (1987). "A quasilinear parabolic system arising in modeling of catalytic reactors," *Journal of Differential Equations* **70**, pp. 167–196.

Gilbarg, D. and Trudinger, N. S. (1998). *Elliptic Partial Differential Equations of Second Order* (Springer, Berlin, Germany).

Griso, G. (2004). "Error estimate and unfolding for periodic homogenization," *Asymptotic Analysis* **40**, pp. 269–286.

Gurtin, M. (1993). *Thermomechanics of Evolving Phase Boundaries in the Plane* (Oxford University Press, Oxford).

Gustafsson, B. and Mossino, J. (2003). "Non-periodic explicit homogenization and reduction of dimension: The linear case," *IMA Journal of Applied Mathematics* **68**, 3, pp. 269–298.

Heida, M. (2011). "An extension of the stochastic two-scale convergence method and application," *Asymptotic Analysis* **72**, 1-2, pp. 1–30.

Hilhorst, D., Martin, S., and Mimura, M. (2012). "Singular limit of a competition–diffusion system with large interspecific interaction," *Journal of Mathematical Analysis and Applications* **390**, 2, pp. 488–513.

Holmbom, A. (1997). "Homogenization of parabolic equations an alternative approach and some corrector-type results," Applications of Mathematics **42**, pp. 321–343.

Holmes, M. H. (1995). *Introduction to Perturbation Methods*, Texts in Applied Mathematics, Vol. 20 (Springer Verlag, Berlin).

Hornung, U. (ed.) (1997). *Homogenization and Porous Media*, Interdisciplinary Applied Mathematics, Vol. 6 (Springer-Verlag, New York).

Hornung, U. and Jäger, W. (1991). "Diffusion, convection, adsorption, and reaction of chemicals in porous media," *Journal of Differential Equations* **92**, 2, pp. 199–225.

Hornung, U., Jäger, W., and Mikelić, A. (1994). "Reactive transport through an array of cells with semi-permeable membranes," *RAIRO Modélisation Mathématique et Analyse Numérique* **28**, 1, pp. 59–94.

Höpker, M. (2016). *Extension Operators for Sobolev Spaces on Periodic Domains, Their Applications, and Homogenization of a Phase Field Model for Phase Transitions in Porous Media*, PhD thesis (University of Bremen, Germany).

Jävergård, N., Morale, D., Rui, G., Muntean, A., and Ugolini, S. (2025). "A hybrid model of sulphation reactions: stochastic particles in a random continuum environment," arXiv:2503.01856 [physics.chem-ph], https://arxiv.org/abs/2503.01856.

Knoch, J., Gahn, M., Neuss-Radu, M., and Neuss, N. (2023). "Multi-scale modeling and simulation of transport processes in an elastically deformable perforated medium," *Transport in Porous Media* **147**, pp. 93–123.

Korhonen, T., Lagerlöf, J. H., and A. Muntean (2020). "Computational study of the effect of hypoxia on cancer response to radiation treatment," *ROMAI Journal* **16**, 2, pp. 75–86.

Kumazaki, K. and Muntean, A. (2025). "A two-scale model describing swelling in porous materials with elongated internal structures," *Quarterly of Applied Mathematics* **83**, pp. 507–532 doi:https://doi.org/10.1090/qam/1705.

Lakkis, O., Muntean, A., Richardson, O., and Venkataraman, C. (2024). "Parallel two-scale finite element implementation of a system with varying microstructure," *GAMM Mitteilungen,* **47**, 4, p. e202470005.

Larsson, S. and Thomée, V. (2003). *Partial Differential Equations with Numerical Methods*, Texts in Applied Mathematics, Vol. 45 (Springer, Berlin, Heidelberg).

Le Bris, C. (2015). *Systèmes multi-échelles*, Mathématiques et Applications (SMAI), Vol. 47 (Springer, Cham, Switzerland).

Leguillon, D. (1997). "Comparison of matched asymptotics, multiple scalings and averages in homogenization of periodic structures," *M3AS* **7**, pp. 663–680.

Lind, M., Muntean, A., and Richardson, O. (2020). "A semidiscrete Galerkin scheme for a coupled two-scale elliptic–parabolic system: well-posedness and convergence approximation rates," *BIT Numerical Mathematics* **60**, pp. 999–1031.

Lukkassen, D., Nguetseng, G., and Wall, P. (2002). "Two-scale convergence," *International Journal of Pure and Applied Mathematics* **2**, 1, pp. 35–86.

Lukkassen, D. and Wall, P. (2005). "Two-scale convergence with respect to measures and homogenization of monotone operators," *Journal of Function Spaces and Applications* **3**, pp. 125–161.

Marchenko, V. A. and Khruslov, E. Y. (2006). *Homogenization of Partial Differential Equations* (Birkhäuser, Basel, Switzerland).

Marciniak-Czochra, A. and Ptashnyk, M. (2008). "Derivation of a macroscopic receptor-based model using homogenization techniques," *SIAM Journal on Mathematical Analysis* **40**, 1, pp. 215–237.

Marschoun, L. T., Muthukumarappan, K., and Gunasekaran, S. (2001). "Thermal properties of Cheddar cheese: experimental and modeling," *International Journal of Food Properties* **4**, 3, pp. 383–403.

Mei, C. C. and Vernescu, B. (2010). *Homogenization Methods for Multiscale Mechanics* (World Scientific, New Jersey).

Meier, S. (2008). *Two-scale models for reactive transport and evolving microstructure*, Ph.D. thesis (University of Bremen, Germany).

Meier, S. A. and Muntean, A. (2008). "A two-scale reaction–diffusion system with micro-cell reaction concentrated on a free boundary," *Comptes Rendus Mécanique* **336**, pp. 481–486.

Meier, S. A. and Muntean, A. (2010). "A two-scale reaction-diffusion system: Homogenization and fast reaction limits." in *Current Advances in Nonlinear Analysis and Related Topics, Gakuto International Series: Mathematical Sciences and Applications*, Vol. 32 (Gakkotosho, Tokyo, Japan), pp. 443–461.

Meier, S. A., Peter, M. A., Muntean, A., Böhm, M., and Kropp, J. (2007). "A two-scale approach to concrete carbonation," in *Proceedings of the International RILEM Workshop on Integral Service Life Modeling of Concrete Structures* (Guimares, Portugal), pp. 3–10.

Ming, P. and Song, S. (2024). "Error estimate of multiscale finite element method for periodic media revisited," *Multiscale Modeling and Simulation* **22**, 1, pp. 106–124.

Müller, S. (1999). *Variational models for microstructure and phase transitions*, Vol. 1713, Chap. 1 (Springer Berlin Heidelberg, Berlin, Heidelberg), pp. 85–210.

Muntean, A. (2015). *Continuum Modeling: An Approach Through Practical Examples*, SpringerBriefs in Mathematical Methods (Springer, Cham, Switzerland).

Muntean, A. (2023). "Multiscale carbonation reactions: Status of things and two modeling exercises related to cultural heritage," in G. Bretti, C. Cavaterra, M. Solci, and M. Spagnuolo (eds.), *Mathematical Modeling in Cultural Heritage* (Springer Nature, Singapore), pp. 175–185.

Muntean, A. and Chalupecky, V. (2011). *The Homogenization Method and Multiscale Modeling* (Kyushu University, Japan).

Muntean, A. and Neuss-Radu, M. (2010). "A multiscale Galerkin approach for a class of nonlinear coupled reaction-diffusion systems in complex media," *Journal of Mathematical Analysis and Applications* **371**, 2, pp. 705–718.

Muntean, A. and Nikolopoulos, C. (2020). "Colloidal transport in locally periodic evolving porous media—an upscaling exercise," *SIAM Journal on Applied Mathematics* **80**, 1, pp. 448–475.

Muntean, A. and van Noorden, T. (2013). "Corrector estimates for the homogenization of a locally periodic medium with areas of low and high diffusivity," *European Journal of Applied Mathematics* **24**, 5, pp. 657–677.

Murad, M. A. and Cushman, J. H. (1996). "Multiscale flow and deformation in hydrophilic swelling porous media," *International Journal of Engineering Science* **34**, pp. 313–338.

Nepal, S., Raveendran, V., Eden, M., Lyons, R., and Muntean, A. (2024). "Numerical exploration of nonlinear dispersion effects via a strongly coupled two-scale system," arXiv:2402.09607v2.

Neuss-Radu, M. (1992). *Homogenization techniques*, Diploma thesis (University of Cluj-Napoca, Romania, and University of Heidelberg, Germany).

Neuss-Radu, M. (1996). "Some extensions of two-scale convergence," *Comptes Rendus de l'Académie des Sciences, Série I: Mathématique* **332**, pp. 899–904.

Nguetseng, G. (1989). "A general convergence result for a functional related to the theory of homogenization," *SIAM Journal on Mathematical Analysis* **20**, pp. 608–623.

Nika, G. and Muntean, A. (2023). "Hypertemperature effects in heterogeneous media and thermal flux at small-length scales," *Networks and Heterogeneous Media* **18**, 3, pp. 1207–1225.

Nishiura, Y. (2002). *Far-from-Equilibrium Dynamics* (American Mathematical Society, Providence, RI).

Ozdilek, E. E., Ozcakar, E., Muhtaroglu, N., Simsek, U., Gulcan, O., and Sendur, G. K. (2024). "A finite element based homogenization code in Python: HomPy," *Advances in Engineering Software* **194**, p. 103674.

Panasenko, G. P. (2002). "Partial homogenization," *Comptes Rendus: Mécanique* **330**, 10, pp. 667–672.

Pavliotis, G. A. and Stuart, A. M. (2008). *Multiscale Methods: Averaging and Homogenization*, Texts in Applied Mathematics, Vol. 53 (Springer, New York).

Pawlowski, J. (1971). *Die Änlichkeitstheorie in der physikalisch-technischen Forschung. Grundlagen und Anwendung* (Springer Verlag, Berlin).

Pedregal, P. (2005). "Multi-scale Young measures," *Transactions of the American Mathematical Society* **358**, 2, pp. 591–602.

Persson, L. E., Persson, L., Svanstedt, N., and Wyller, J. (1993). *The Homogenization Method: An Introduction*, Studentlitteratur (Chartwell Bratt, Lund, Sweden).

Peter, M. A. and Böhm, M. (2008). "Different choices of scaling in homogenization of diffusion and interfacial exchange in a porous medium," *Mathematical Methods in the Applied Sciences* **31**, 11, pp. 1257–1282.

Presutti, E. (2009). *Scaling Limits in Statistical Mechanics and Microstructures in Continuum Mechanics*, Theoretical and Mathematical Physics (Springer, Cham, Switzerland).

Ptashnyk, M. (2013). "Two-scale convergence for locally periodic microstructures and homogenization of plywood structures," *Multiscale Modeling and Simulation* **11**, 1, pp. 92–117.

Raveendran, V., Cirillo, E. N. M., de Bonis, I., and Muntean, A. (2022). "Scaling effects on the periodic homogenization of a reaction-diffusion-convection problem posed in homogeneous domains connected by a thin composite layer," *Quarterly of Applied Mathematics* **80**, pp. 157–200.

Raveendran, V., Nepal, S., Lyons, R., Eden, M., and Muntean, A. (2025). "Strongly coupled two-scale system with nonlinear dispersion: Weak solvability and numerical simulation," *ZAMP*.

Renardy, M. and Rogers, R. C. (2004). *An Introduction to Partial Differential Equations*, Texts in Applied Mathematics, Vol. 13 (Springer Verlag, New York).

Richardson, O., Jalba, A., and Muntean, A. (2019). "Effects of environment knowledge in evacuation scenarios involving fire and smoke: A multiscale modelling and simulation approach," *Fire Technology* **55**, pp. 415–436.

Saloff-Coste, L. (2002). *Aspects of Sobolev-type Inequalities*, Lecture Note Series of London Mathematical Society, Vol. 289 (Cambridge University Press, Cambridge).

Sanchez-Hubert, J. and Sanchez-Palencia, E. (1992). *Introduction aux méthodes asymptotiques et á l'homogénéisation: Application à la mécanique des milieux continus*, Collection Mathematiques appliquées pour la maitrise (Masson, Paris).

Sanchez-Hubert, J. and Sanchez-Palencia, E. (1993). *Exercises sur les méthodes asymptotiques et l'homogénéisation*, Collection Mathematiques appliquées pour la maitrise (Masson, Paris).

Seidman, T. I. (2009). "Interface conditions for a singular reaction-diffusion system," *Discrete Continuous Dynamical Systems - S* **2**, pp. 631–643.

Showalter, R. E. (1997). *Monotone Operators in Banach Space and Nonlinear Partial Differential Equations* (American Mathematical Society, Providence, RI).

Soyarslan, C. and Pradas, M. (2024). "Physics-informed machine learning in asymptotic homogenization of elliptic equations," *Computer Methods in Applied Mechanics and Engineering* **427**, p. 117043, doi:https://doi.org/10.1016/j.cma.2024.117043, https://www.sciencedirect.com/science/article/pii/S0045782524002998.

Tartar, L. (1990). "Memory effects and homogenization," *Archives for Rational Mechanics and Analysis* **111**, pp. 121–132.

van Duijn, C. J. and Pop, I. S. (2004). "Crystal dissolution and precipitation in porous media: Pore scale analysis," *Journal für die reine und angewandte Mathematik* **2004**, 577, pp. 171–211.

van Lith, B. S., Muntean, A., and Storm, C. (2014). "A continuum model for hierarchical fibril assembly," *Europhysics Letters* **106**, 6, p. 68004.

van Meurs, P., Muntean, A., and Peletier, M. (2014). "Upscaling of dislocation walls in finite domains," *European Journal of Applied Mathematics* **25**, 6, pp. 749–781.

van Noorden, T. L. and Pop, I. S. (2008). "A Stefan problem modelling crystal dissolution and precipitation," *IMA Journal of Applied Mathematics* **73**, 2.

Visintin, A. (2006). "Towards a two-scale calculus," *ESAIM: Control, Optimisation and Calculus of Variations* **12**, 3, pp. 371–397.

Vo Anh, K. and Muntean, A. (2019). "Corrector homogenization estimates for a non-stationary Stokes-Nernst-Planck-Poisson system in perforated domains," *Communications in Mathematical Sciences* **17**, pp. 705–738.

Vromans, A. J., van de Ven, F., and Muntean, A. (2019). "Homogenization of a pseudo-parabolic system via a spatial-temporal decoupling: Upscaling and corrector estimates for perforated domains," *Mathematics in Engineering* **1**, 3, pp. 548–582.

Weinan, E., (2011). *Principles of Multiscale Modeling* (Cambridge University Press, Cambridge).

Wolf, S. (2022). *Some Homogenization Problems in a Periodic Setting with a Local Defect*, Ph.D. thesis (Université Paris Cité, Paris).

Zeidler, E. (1990). *Nonlinear Functional Analysis and its Applications: Linear Monotone Operators*, Vol. II/A (Springer, Berlin, Heidelberg).

Zhikov, V. V. and Piatnitskii, A. L. (2006). "Homogenization of random singular structures and random measures," *Izvestiya: Mathematics* **1**, pp. 19–67.

Index

P

R

S

T

U

Y